Creating Science!

Hands-on Science Activities & Experiments for Everyone!

National Library of Australia Cataloguing-in-Publication entry Author: Ireland, Joe.

Title: Creating Science - Hands-on Science Activities & Experiments for Everyone!

Author: Dr Joseph Ireland "Dr Joe".

Target Audience: For primary school age.

Subjects: Science / Experiments & Projects.

Dewey Number: 507.8

ISBN: 978-0-9923294-19   8.5" x 11"(21.59 x 27.94 cm)

For my dad, who taught me to love science,

and my mum, who taught me it's OK to have questions.

With extra thanks to Grandpa Norm, who showed me how to make fun technological toys from old recycled materials and then, after his stroke, just how wonderful and important a story can be.

Editorial gratitude to Kathy Dash, Brisbane Southside Homeschoolers, Sunnycoast North Homeschoolers, West End State School, Clinton OSHC, the talented Carl Antuar, and of course with infinite gratitude to all participants!

Text and attributions

Attributions - all art and pictures were by the publisher during home science, or with the help and permission of various school and home school students in Australia, for which we are deeply indebted. All other media are available freely online, but should not be taken to support this creative work or any assertions made herein.

Picture of famous scientists taken 17 April 2017 from: **Ruby Payne-Scott** at https://en.wikipedia.org/wiki/File:Peter-hall.ruby.payne-scott.jpg **Barry Marshall** By Barjammar - Own work, Public Domain, https://commons.wikimedia.org/w/index.php?curid=13167242 **Dr John Cade** http://www.abc.net.au/news/2016-10-29/dr-john-cade-developed-lithium-to-treat-mental-illnesses/7962438

Andromeda Galaxy, in "gravity", taken 6th August 2016 - By Adam Evans [CC BY 2.0 (http://creativecommons.org/licences/by/2.0)], via Wikimedia Commons

Emu, Toucan, and Penguin by **Emily Ireland**, 2018. Dr Joe Chibi's by **Karlie Ireland**, 2018. Wind cannon photo's by family members LC and Samantha Ireland.

Contents

How To Make Science Happen

Question and Predict

What would you like to learn more about? You're the scientist now!

From what you know and have experienced, gather information and in order to think of ideas that explain how the world works. Use those ideas to make predictions about what will happen.

Communicate

Science must be shared! Consider your evidence, justify your position and present your conclusions. Can you present your finding to others to convince them of your logic and evidence? Answer the WHAT and they WHY questions.

Dr. Joe
the Travelling Scientist, PhD

Proudly presented by **Creating Science**

Plan and Conduct

Think of ways to test your ideas. Collaborate with others, and make sure you have all the equipment you need, and make a plan for recording your results.

Now: Test. Remember – safety first!

Evaluate

Compare your thinking, results and methods with others. What did you learn about the world? What did you learn about learning through science?

What might work better next time?

Process and Analyse

What are your results trying to teach you? What can you do to improve the accuracy of your conclusions? Remember *experiments always work* – because their job is to teach us something new, not just to entertain us.

So many different kinds of science using the same method!

REMEMBER: SCIENCE IS DANGEROUS!

Parental Supervision is Required!

Not just for safety - science is fun!

This book is designed to be used by kids and grownups *working together*; safely, and responsibly. In using this book you're promising to take care of your little scientist, helping them to stay safe and manage the dangers of science.

If you'd like to set up a home lab, here's a few things you'll need:

Safety: A clean space, hard floor or drop-sheet over carpet. Old clothes. Covered shoes. Eye covering. Long hair tied back. Gloves. Appropriate fire safety plan.

Art supplies: Scissors. PVA glue. Craft materials including cardboard boxes and tubes, feathers, seed pods, pipe cleaners, etc. Sticky tape. Blu Tack.

Parents' help: Hot melt glue gun, Stanley knife, Cooking equipment and skills. Fire extinguisher. Supervision and companionship.

Let's share that fun with others!

Find us anywhere online with the hashtag

#CreatingScience

Followed by the chapter title.

For example, #CreatingScienceBobTheBlob or #CreatingScienceRocketBalloons.

Share experiences, Give feedback, Find and offer helpful suggestions. #CreatingScienceWithDrJoe

Creating Science is for everyone!

Science doesn't just drop out of a tree and onto our heads!

We have to *make* science. Who do you think came up with all these wonderful ideas about how the world works? It was people, like you and me.

So how did we do it? Are we still doing it? And, perhaps m o s t importantly, is creating s c i e n t i f i c k n o w l e d g e s o m e t h i n g *you* can do as well?

I would answer with a definite YES! Whether you become a scientist or not, you will make decisions based on what people call science - including what you buy and who you vote for. You need to know what science is, what it isn't, and how it is created so that it can help you make the best decisions in life.

So what does it take to create science? Well, grab a grownup and try out some of the experiments and activities in this book, and you'll begin to get an idea not only of how we see the world through science, but how you can help to create the knowledge that will make it a better world for everyone!

In this book you will learn how to be a scientist through the following, and you can try out some related activities:

- Science begins with **questions** which we will learn about (via Puff Bottles and Air Pressure).
- Questions are answered using **theories** (via Bob the Blob and the science of buoyancy).
- T h e o r i e s are tested using **experiments**, which often use *variables* (via Rocket Balloons). Experiments also make use of *multiple trials* and *fair testing* (Reaction Time) and *exact measures*, (Bounce Master), etc.
- We scientists must **conclude** what those experiments mean (via The Returning Roller) and evaluate our work.
- We then **communicate** our conclusions (learn about this via the Magician's Steal).

Of course, there's more to it than that! But this'll do to get us started...

- Science affects us. It influences almost every decision we make in this modern world. Science is tied up with technology and society. It is changing the way we live every day. (Via the Science of Spectrometry, and much, much more!)
- Scientific ideas always belong to *someone*. Those people have their own cultures, and sometimes live in very different times. Who made up the idea, when, and why? (Via Science is People.)

Advanced stuff

Research shows that students need to know more about what science is in order to be able to do, and enjoy, science. We call these the virtues of science. These ideas are so important that they are interwoven with every science activity in this book.

- Science is **Creative**. Whether you're trying to explain something new, or trying to think of a way to test that explanation, you're going to need to be very creative to be a scientist! It's one of the most *fun* things about it!

- Science is **Tentative** (we're still looking for a better word than this). Science is *allowed* to be *wrong*. We change our minds when better ideas, or better evidence, arrive. It's one of the most *powerful* things about science as a way of knowing.

- Science is also **Empirical**. You know, it doesn't (or at least, shouldn't) matter if a scientist is rich, famous, or even a world leading expert. What matters is: can their ideas be tested, and are their ideas supported by experiments? This sets science apart from many other ways of knowing.

And we're not even close to finished. Meet the values of science!

- Science values **Objectivity**. While recognising that it's not always possible, and sometimes embracing subjectivity as part of the research, science strives to explain things as they *are*, not as we'd *like* them to be. It's not easy!

- **Replicability.** Perhaps the most important value of science is to create knowledge that we're so sure of, *anyone else* should be able to use our work and get the same results.

- Science values **Honesty**. We trust that other scientists aren't "fudging their data," and making up results just to support the ideas they like.

- Science values **precision**. Scientists like to be exact, and we like theories that work exactly.

- **Collaboration** is vital in science. We work in teams, and we often need to share our science with someone.

- Science values **consensus**. When the majority of scientists working in a field agree on a certain idea, we say it has achieved *consensus*. Some say science cannot tell when a theory is *true*, but we do know when scientists agree it *works*.

- Most scientists embrace **realism:** The idea that there is a real world, and that we can know that world, and explain and predict that world using our ideas and words.

And what's life without a few definitions to keep us alive!

- **Theory** means *testable explanation*. **It does not mean 'guess'.** It is usually a complex explanation, supported very well by available evidence, and protected by scientific consensus. We use theories to understand the world and make predictions. Theories always start with the word 'because'. They are often answers to the question 'how come?'

- **Prediction** means a *guess of what will happen*. We use this word every day and in science. Scientific theories often assume things which cannot be observed directly, so we don't test theories themselves, we test their predictions. Thus many scientists believe theories are never wrong or right, they are either supported, or not supported by evidence.

- **Hypothesis** means *new theory* (though you can be more complex than that if you'd like). The word hypothesis does NOT mean

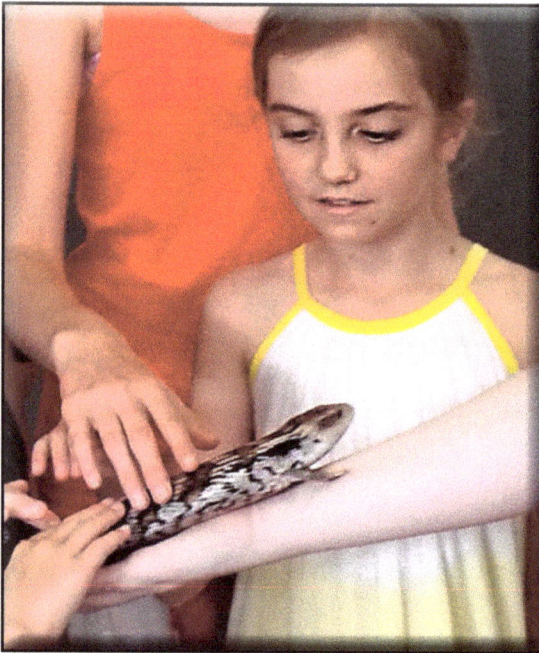

- Science uses **experiments** to explore ideas about the world. Experiments *never* fail - they always tell us something about the world (even if that something is 'never do that again!'). Experiments begin, and end with, questions. Remember folks - experiments are *not* the ONLY way we learn about the world as scientists! There is no one 'scientific method'.

- Science also makes good use of **demonstrations**. Most science in schools, and many in this book, are not real experiments to *test explanations* (aka, theories), they are activities we do to *prove a point*.

- **Scientific Laws** are another kind of theory. They explain a clear relationship in the world, usually mathematically. Theories do not grow into laws! Laws are just another explanation!

- **Observations** are what we sense; things we see or hear, for example. **Inference** is what those sensations mean. We can't feel the earth move, but we infer day and night occur because it does. Knowing the difference is vital.

prediction. Hypotheses usually explain only a few observations, and might help us to develop a complete theoretical framework one day. Social sciences would do well to heed this advice.

Need a more in depth discussion on the history of philosophy of science and what it means to you personally? Stay tuned for my upcoming Textbook "Creating Science - The Textbook"

Science is a people activity

Who invented a theory matters; what's their background, motives, experience?

Every science idea ever created was made by a person – who?

How they say it matters. What evidence do they have?

Science ideas must be tested before being accepted.

What are they actually saying, and what does it mean?

What did they teach us? Was it electricity, motion?

Can this new knowledge be used to predict and work with nature?

As inspired by the Australian national curriculum:

Human Influences (Who)

Science Inquiry Skills (How)

Science Content (What)

Creating Science!

Science Inquiry Skills

Puff Bottles

INQUIRY SCIENCE

Science begins with questions, and the direction science will take in the future will depend very much on the kinds of questions you're going to ask. Try this activity to get your friends puzzling away!

Safety

Remember to be healthy by **not sharing your balloon** with others – they can use their own balloon or build their own bottle! Also, make sure you **get a grown up to drill the hole** in the bottom of the bottle that you'll need.

Materials:

- Empty soft drink bottles (lids are not required). Small tops and wide sides, are preferred, with labels removed. Strong, plastic bottles are best - larger ones tend to crumple.

- You'll need a grown up to drill a small hole in the bottom of the bottle – as large as possible, but still small enough to cover with your finger (3-4mm is usually best).

- A balloon large enough to cover the opening of the bottle. (Water balloons won't do – unless you have a *very* small bottle!)

#CreatingSciencePuffBottles

Building Puff Bottles:

- Gather the materials.

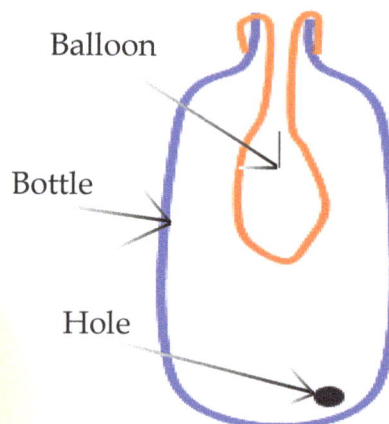

Balloon
Bottle
Hole

- It is helpful if you stretch your balloon out several times first. This makes it easier to blow them up.

- Push the balloon into the bottle and pull the end back over the lid (see the picture). Be sure the balloon seals around the entire top so that no air can escape around the edge.

Using the Puff Bottles:

A little bit of magic science can go a long way! Blow up the balloon. Then, while you click or tap with one hand you also hold the hole closed with your other hand's finger – and the balloon stays up! Wait as long as you like and when you're ready, click again and remove your finger – the balloon will 'magically' deflate! Can your friends and family work out how you do it?

*Remember – finger off to inflate, finger on to keep the balloon inflated… with the **magic of science!***

Asking questions is the FIRST STEP in science!

Why it Works

There are many very good explanations of why this works – here's mine: all air everywhere is pushing all the time – we call it air pressure and it happens because there is a lot of air between you and outer space. It's pushing down and around you. (And it's not pushing lightly – it's pushing very hard!)

Explain More!

When you blow up the balloon you push the air out of the lower part of the bottle. But then you cover the hole with your finger preventing any air from pushing its way back into the bottle *except* through the balloon which is still open to the outside air – and it pushes the balloon in as far as it can go until it balances the air pressure inside the bottle. Release the hole, and the air can push into the bottle from both sides of the balloon. As the air pressure evens out the balloon (which was pulled tight like an elastic band) returns to its preferred shape – an uninflated balloon!

To help us understand this better, we'll need to explore an idea called pressure, which is a foundational concept we'll come back to in other topics like space, weather, etc!

INQUIRY SCIENCE

Air Pressure

Air pressure is one of the easiest concepts to understand, but one of the hardest to believe!

INQUIRY SCIENCE

1. Is air weak? Try to *squash* a clean, empty soft drink bottle. Notice how it feels to squish the bottle?

2. Now put the lid on the bottle, is it just as easy to squash the bottle now?

3. Why it is harder to squash the bottle with the lid on? Can *anyone* squash the air?

1. Scrunch some paper into the bottom of a cup.

2. Put the cup upside down in water.

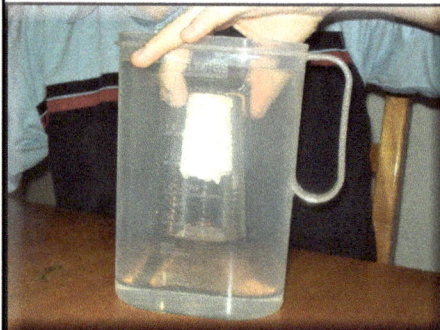

3. Why doesn't the water come into the cup?   The paper will even stay dry!

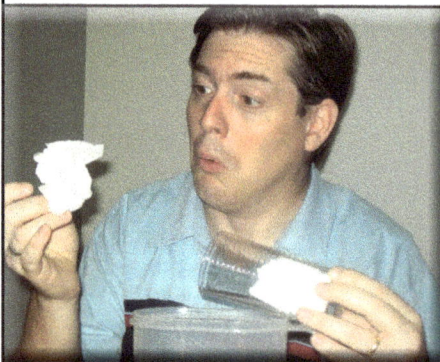

1. Is air strong enough to hold up a person? Blow up enough balloons to hold up an inverted table.

2. Have someone stabilise the table while someone else stands on it carefully!

3. Why don't the balloons all squish flat, even with a heavy grownup?

The answer to all these questions is that *air has pressure*. Imagine the air as billions of tiny bouncing particles. These particles are always bouncing and never get tired. Even though the individual particles are very small, all of them bouncing together create enormous **pressure**. The pressure created by ordinary air is enough to keep water out of a cup, prevent a bottle from being crushed, or hold a grown up off the ground! This idea, called air pressure, helps to explain many of the properties air has.

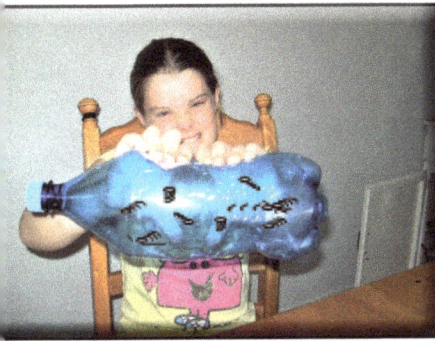

Air pushes the water out!

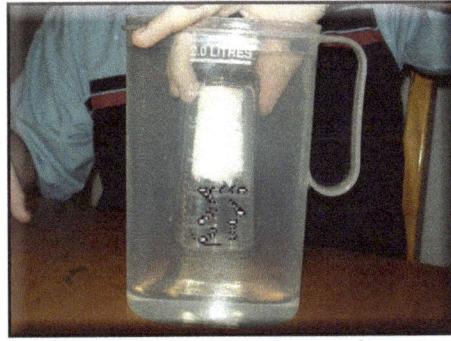

Air stops the bottle crushing!

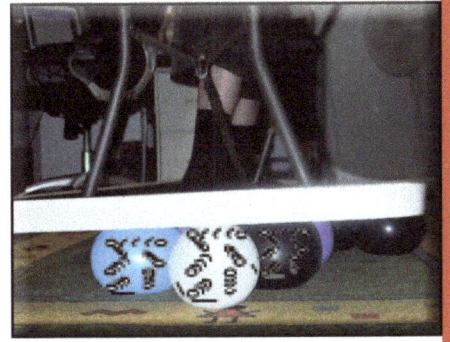

Air can hold you up!

Try some more fun with air pressure!

#CreatingScienceUndrinkableBottle

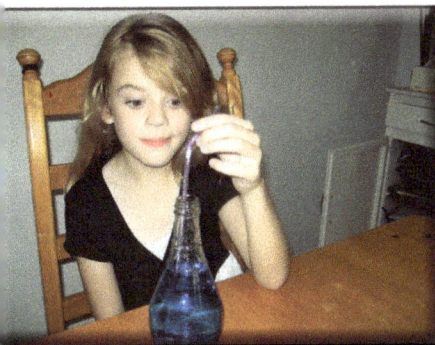

1. Get yourself a glass drink bottle with a thin opening and a lot of something to drink inside.
2. Thoroughly block off the opening (but not the straw!) using Blu Tack or similar.
3. Try and drink though the straw - can you? Air pushes the drink into your mouth!

#CreatingScience PeeingCan

Check this out! I have a pee-on-command can! It's leaking water right now, but when I click my fingers... it stops!

How is that possible?!?!

Easy! The top and bottom have a hole in them. When I click, I cover the top hole, and after a second, water won't come out the bottom either... But good science tends to create more questions for every question that it answers: so why does covering the top hole stop water coming out the bottom?!?

Is it the air pressure pushing on the water underneath, preventing it from leaving when there's not enough air pressure from above?

Here's how to make a pee-on-command cup: have a grown up punch two holes into the top and bottom of an old, clean, can. Fill with water and enjoy - finger on top, water stays in. Finger off, water pees out - why!?!!

Trickier science! Air particles don't weigh very much, so we can push them away with our hands, but there's so many of them that all together they make an enormous push – almost a whole kilogram on a space the size of your fingernail! That's about a ton of pressure over your whole body, almost as much as a car! You can hardly even feel it though because it's always been there. The higher up you go the less air there is, so there is less pressure.

INQUIRY SCIENCE

Generate Theories with...

Bob the Blob

I N Q U I R Y   S C I E N C E

Buoyancy - what an amazing idea! But like air pressure, some science ideas are tricky! Still, if they work, we have to use them!

Materials

- An empty, see-through, clean and dry soft drink bottle (with a lid).
- A modelling balloon, like what clowns use.
- Some Blu Tack (or similar pliable and also water durable product).
- A large cup (filled with water) – see through cups are very useful.
- Water, and probably a kitchen sink – be sure to watch every drop!

How to make Bob the Blob

First, you need to make your diver Bob: but to get it right, he'll need to have just the right balance of air inside and weight underneath.

1. Tie a small bubble of air inside the end of the balloon and cut off the ends. Balloons like to float anyway, so the bubble should be small.

2. Wrap a glob of Blu Tack around the knot, usually half the size of Bob. It needs to be on secure and tight.

3. This is your Bob. Float him in the water filled cup: If he sinks straight to the bottom, take some Blu Tack off. If he floats right out of the top, add more (see diagram below).

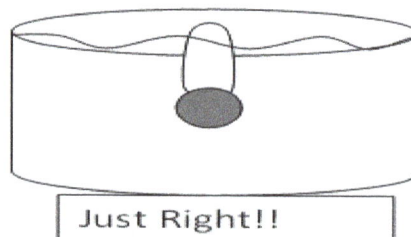

4. Next, fill the bottle with water right up to the very top, so that it's spilling out. Avoid air bubbles because it makes it more difficult to get Bob to work.

5. Drop Bob gently into the bottle. Screw the bottle lid on tightly. Clean up excess water.

And there you have it, your very own Bob the Blob!

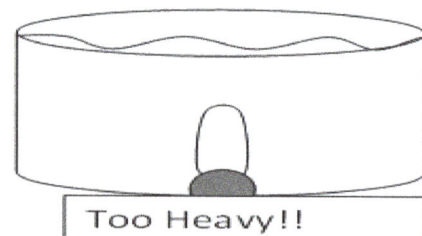

Too Light!!

Just Right!!

Too Heavy!!

   (c) Dr Joseph Ireland 2018

In Science, it's ALL theories!

Buoyancy Rules!

Theories

Yes - lovely, helpful theories. And the best thing about science theories is that they can all be tested! Can your friends figure out what really makes Bob swim down? Is it the sound waves? A matter of timing? Or maybe something completely different?

Remember: In Science, theory means 'testable explanation' not 'guess'!

Of course, they'll figure out you're just squeezing the bottle eventually.

But just as science begins with questions, it ends with them as well! For example: Why does squeezing the bottle make Bob the Blob float downwards? Our science answer is called "buoyancy!"

Squeeze the Bottle

What happens to Bob?

You can play a trick on your friends – tell your friends that Bob is really alive, and you can prove it! Try clicking while you secretly squeeze the bottle with your other hand, and Bob will swim down anyway. Is he alive? Or is this just a trick…?

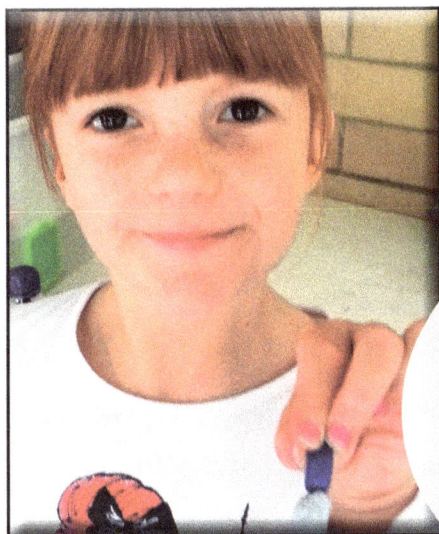

Who Came Up With Buoyancy?

Much of the credit goes to the ancient scientist Archimedes. You know, the one who was trying to find out if the king's crown was made of pure gold? Do you know how he did it?

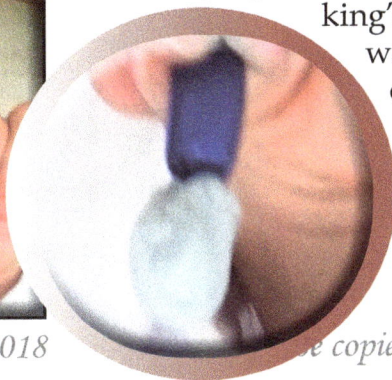

Buoyancy and Bob:

1.   The water is trying to push Bob the Blob up. Ever noticed how things feel lighter in water? Water always pushes things up: the more space they take up, the more the water tries to lift them. You can feel this by trying to push a balloon full of air into a bucket of water. The more water you push out of the way, the harder it gets to push the balloon down! This is buoyancy.

2.   However, Bob's own weight is pulling him down (due to gravity).

3.   When you squeeze, the water squashes the bubble of air in Bob's head. This makes Bob **smaller**, even though he still *weighs* the same. So he sinks in the water. Relieve the pressure, he gets bigger again, and the water's constant push up is strong enough to lift him again.

Remember, Bob is an amphibian, not aquatic. So take him out when not in use or he will fill with water and drown!

Who will win, gravity or water pressure? You decide!

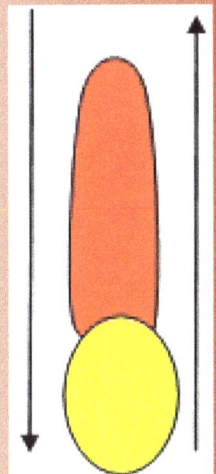

INQUIRY SCIENCE

Rocket Balloons

An important part of Creating Science through experiments: Are we actually testing what we think we're testing? An important way to do that is to make sure we use fair testing.

INQUIRY SCIENCE

Explore what factors make a rocket balloon fly the furthest!

You will need

- A long piece of string. Something strong and smooth is best, such as nylon string - not cotton. Tie it up somewhere safe and run along for some distance (~4-8 meters).
- Some straws, wide enough to have the string fit through.
- Balloons. All sizes, shapes and colours are welcome.
- Sticky tape, scissors, and lots of room.

Next you:

- Thread the straw onto the string. Blow up a balloon, and hold the end. Don't tie it; clips can help.
- Tape the balloon to the straw, running along the length of the balloon; from top to nozzle, so the end of the straw points the same way as the balloon.

- Release the balloon and see how far it will propel itself along the string.
- Now experiment: How do you make your balloon go the furthest?

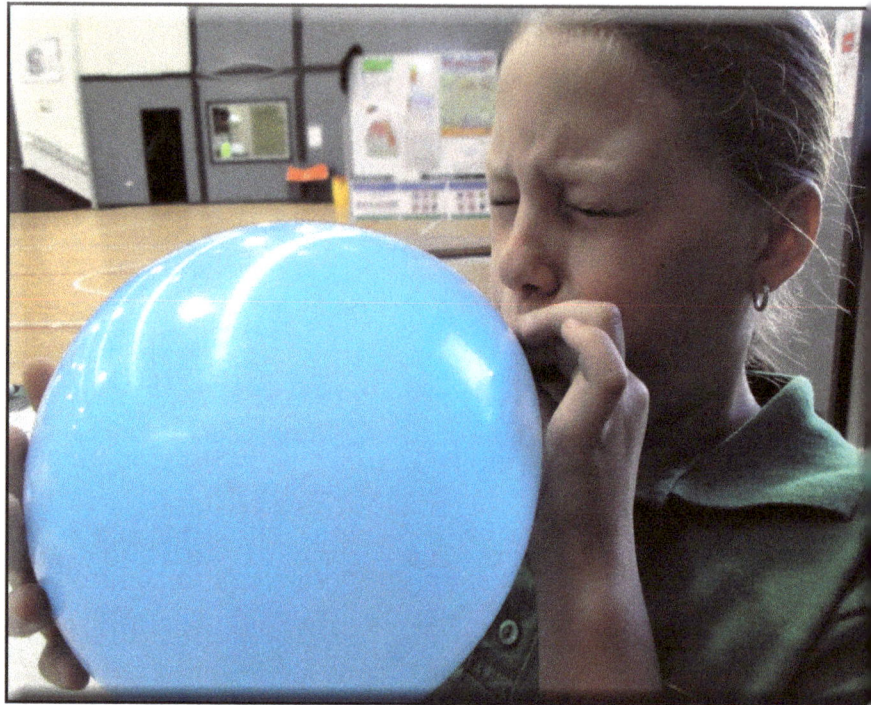

Threading can be tricky!

Experimental Science

In science, the things that affect the results of our experiments, such as how far a balloon will travel along a string, are called **variables**. Explore what variables will help the balloon fly further. You can increase the tension of the string, try different size balloons and alter the length of the straw, for example. The options for possible variables are limitless: string width and friction, attributes of the straw, angle of launch, balloon colours, special nozzles on the end... the list goes on and on...

Remember:

CHANGE only one thing per experiment! Keep all other variables the SAME!!

ANOTHER example of equal and opposite forces: the air is pushing every way - the force that pushes the air out of the balloon also pushes the balloon along the string!!!

How it works

One explanation is that air is pushing all the time, though it may not feel like it! The air inside the balloon is pushing the balloon outwards, helping it to keep its inflated shape. When you open the nozzle the air rushes out, which means the air pushing on the front of the balloon 'wins' and pushes the whole balloon along the string.

Can you make a rocket balloon that will travel the furthest every time?

Theory Testing

So if you increase the weight of the material flowing out of your balloon, will it increase the distance the rocket will go? Might it also increase stability? Or will it simply hold the rocket back?

You can test this by putting some confetti or glitter into your balloon.

Rocket science is pretty simple at heart: If you want something to go right, you have to push something else to the left!

Super Extension

Can you invent a two stage rocket balloon using a modified plastic cup as separator…?

Remember, there's a lot more to experiments than variables alone, and more to science than experiments, this is all just for starters!!!

INQUIRY SCIENCE

Careful, balloons can POP!

Dropping Rulers

We never do an experiment 'at least three times', we experiment as many times as we can!

INQUIRY SCIENCE

Test your reaction time!

You'll need: a normal ruler and someone to help out.

Have your helper hold the ruler as shown above.

Hold your hands down at the other end, not touching the ruler, right at the 0cm mark.

Without warning, at a random moment, have your helper let go of the ruler and see if you can catch it before it falls past!

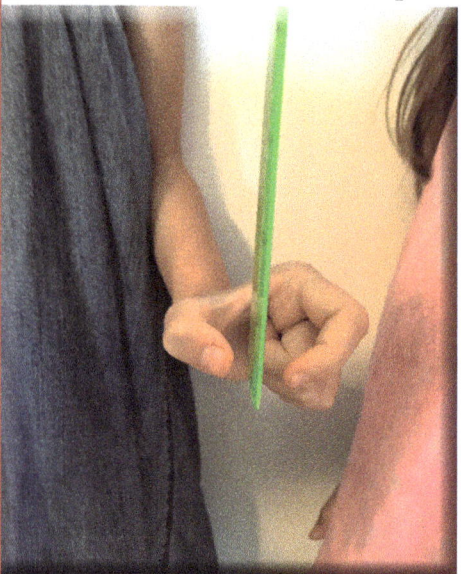

Distance (Cm)	Time (seconds)
1	0.045
2	0.064
3	0.078
4	0.090
5	0.101
6	0.111
7	0.120
8	0.128
9	0.136
10	0.143
11	0.150
12	0.156
13	0.163
14	0.169
15	0.175
16	0.181
17	0.186
18	0.192
19	0.197
20	0.202
21	0.207
22	0.212
23	0.217
24	0.221
25	0.226
26	0.230
27	0.235
28	0.239
29	0.243
30	0.247

Remember, this activity tests more than 'reaction time', it also tests eyesight, hand eye coordination, nerve impulse speed - so if you don't do well, don't worry, this isn't a formal scientific evaluation!

The average is about 20 cm for a grown up - how did you do?

Remember to keep it fair; what variables could mess up the accuracy of the knowledge we are creating?

- Do you blink or give some other indication you're about to let go?
- Are you holding the ruler at different heights?
- Did you let go at a random time, or a predictable time?
- Are you taking so long to let the ruler go, it's more a test of concentration than reaction time?

Now let's get to creating some scientific knowledge!

So what variables affect reaction time? Are boys faster than girls? Are kids full of sugar faster than kids on healthy food - and how about an hour after eating? Would it be **fair** to test reaction time with kids vs adults - adults just after they wake up, and the kids right after a fun game of sports? Not really! Science aims to have *fair* tests;

Keep every variable the same EXCEPT the one we're testing.

This page may be copied for incidental educative purposes only   (c) *Dr Joseph Ireland 2018*

Bounce Master

*Approximations will not do - in science we love to be **accurate!***

Creating the very best knowledge often requires that we be thorough and accurate!

ACTIVITY: PROVE WHICH IS THE BOUNCIEST BALL.

Remember, smashing the ball on the floor won't help. We need *exact measures.*

Using three balls, try to determine the bounce co-efficient of each of them. Use the average of five bounces (or more, if you prefer) to try and get a more accurate answer.

You could try:

1. Tape a 1 meter ruler or tape measure onto the wall.

2. Drop the ball from 1 meter – don't shove, just release.

3. Allow one or more observers to try and measure *exactly* how high the ball bounced back up on its first bounce only (and not the other bounces). The big trick here is to make sure you're accurate! One way is to remember that if you measure 100cm from the bottom of the ball, you MUST measure the bounce height from the bottom of the ball as well, not the top.

Coefficient of Bouncing =

Height of the ball bounce / Height the ball was dropped

4. Record your results carefully.

5. Repeat the experiment as many times as you feel you need to in order to be confident of your results.

6. Take the average of the results in order to try and get a more accurate answer: Add all your results heights' together, then divide that number by the total number of trials you did in order to generate the coefficient of bouncing - AKA bounciness.

7. Repeat the entire experiment from point 1 with at least 3 different balls. Find out which ball was the bounciest.

Science involves testing, but how many tests are best? As many as we can!

*Be **extra accurate** - use a slow-motion camera to record the action with precision!*

If you use maths and it tells you how high the ball will go in one meter, will you be able to tell how high the ball will go in 10 meters?

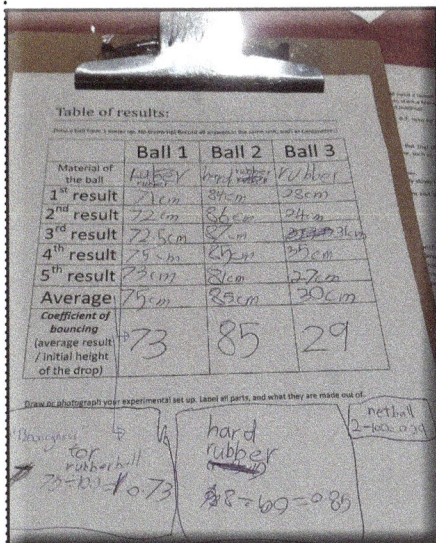

INQUIRY SCIENCE

Table of results:	Ball 1	Ball 2	Ball 3
Material of the ball	rubber	hard rubber	rubber
1st result	71cm	84cm	28cm
2nd result	72cm	86cm	24cm
3rd result	72.5cm	82cm	36cm
4th result	75cm	82cm	34cm
5th result	73cm	81cm	27cm
Average	75cm	85cm	30cm
Coefficient of bouncing (average result / initial height of the drop)	.73	85	29

Returning Roller

To be real science YOU must decide what the results mean; no-one can do it for you! Test your ideas...

Roll a cylinder, and it mystically returns to you every time!

**WARNING:
Tricky project here, grownup help required!**

How to Build a Returning Roller

Caution:

When constructing your returning roller, remember safety first! Be wary of sharp knives and snappy rubber bands. Also, you will need to be persistent; the rollers can be fickle toys (relying on just the right density, weight, & thickness of bands etc.!)

Materials:

- A cylinder at least 10cm wide and 10cm long... or thereabouts. I find large postal rolls do just fine!

- A small, very heavy weight: a nut from a huge bolt, bundle of coins, a solid rock, etc.

- Some strong elastic bands, but not too thick: 1-2mm will usually do.

- Two paperclips or large match sticks.

- Construction materials: sharp knife, tape, etc.

- Perhaps some decoration materials: paper, paints etc.

Is there some trick behind this? We'll have to find out!

#CreatingScienceReturningRoller

Building the Roller

1.   Attach the elastic bands to the weight (being sure to leave enough slack to attach the bands to the cylinder walls in the next step). The bands need to be a little tight, but not too tight, for the best results.

2.   Drill some small slits in the center of either end of the cylinder. Thread the elastic bands through, and keep them there with the paper clips (or match sticks). Your roller should now have a weight suspended by rubber bands inside a cylinder. It is important that the weight does not touch the sides of the cylinder. It's important that the paperclips are stuck to the container or the rubber bands will not transfer their spin properly. Note that sometimes the rubber bands need replacing.

3.   Decorate to your hearts content. Well, nearly. The roller works better with smooth, round walls: I have found cupcake holders are not conducive to this effect.

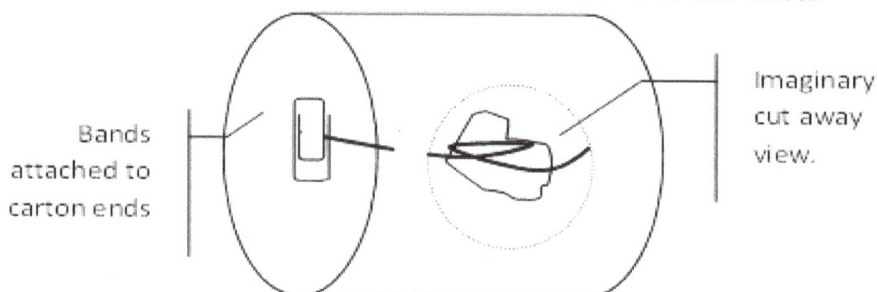

Bands attached to carton ends

Imaginary cut away view.

Help to Make it Work

The roller may require some winding up in the same direction that you intend to roll it (i.e., 'top away from you'). This can be a delicate art: you need to feel the roller just gently trying to roll back. Too much, and the elastic bands unwind by themselves as the weight is spun around inside. Too little and the roller rolls away and is too 'tired' to return. Each roller will be an individual with their own personal requirements to get motivated.

How it Works

This is a chance to show the need for imagination in science. You need to imagine what is happening inside the roller while it is operating.

What happens to the weight as the container is rolling around? Is it turning as the container does? Actually, no! Can you imagine it **not** turning with the container?

What does this do to the rubber band? It causes the elastic band to become

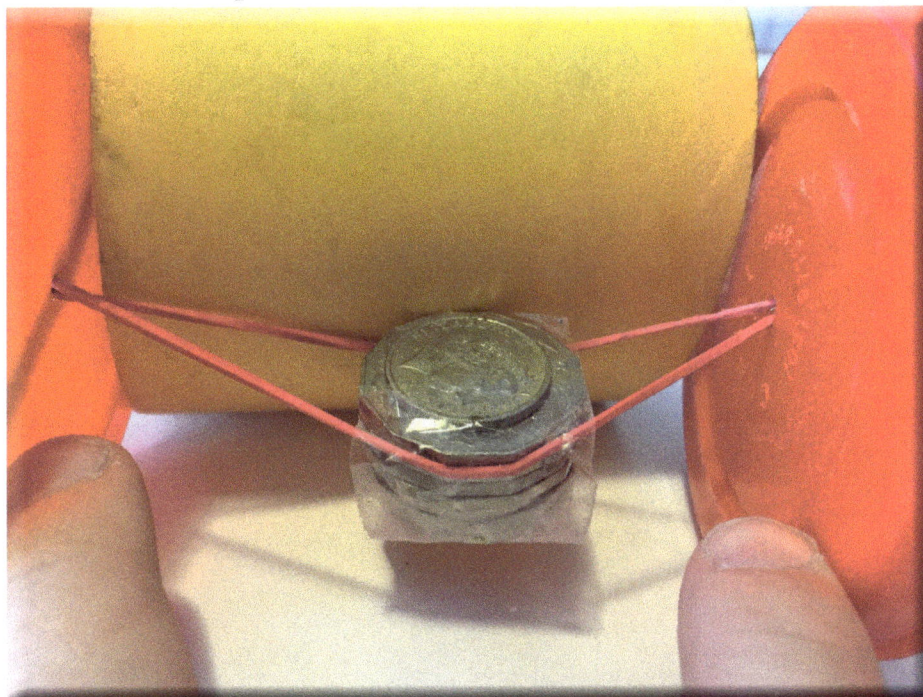

twisted up. This twisting slows the container down as it pulls on the container. Eventually the rubber band begins to untwist and, since its weight is more than the container, it is the container that moves in the opposite direction instead. And presto! The roller returns.

Why it Works

Here's where things get tricky, thus, here's where the science is really!

You can also try describing the returning roller in terms of the scientific concept of **Energy**. The kinetic energy of the original motion is transformed into elastic potential energy in the rubber bands. The tension in the rubber bands produces a force which begins to turn the container in the opposite direction, as the elastic potential energy is transformed back into kinetic energy, and the container rolls back again.

What other examples of forces are there in your classroom? Can you think evidence to support the idea 'moving things keep moving unless acted on by a force' when so many things in our daily lives slow down without constant pushing (cars, bikes, or toys, for instance)? An idea called **friction** will help here.

INQUIRY SCIENCE

Magician's Steal!

INQUIRY SCIENCE

Science must be shared! Can you communicate this trick?

1. Place a small object, such as a coin or rolled up tissue, in your non-dominant hand (for most of us, our left). We find a square piece of foam from an old cut up pillow can be quite helpful. It needs to crush easily, and not be slippery. Make sure your audience can see you placing the object in the palm of your hand. Hold the object there using the pointer finger of your other hand.

2. Close the fingers of your holding hand over your pointing finger. This helps block out your audiences' view for the next step!

#Creating Science Magicians Steal

Can you see through the Science of Magic?

3. Quickly SNATCH the object out of your holding hand, and hold it in the palm of your hand with the pointer finger. With practice, you can't even see yourself do it in a mirror!

NOTE

Make sure you keep your holding hand CLOSED TIGHT or it's not going to trick anyone!

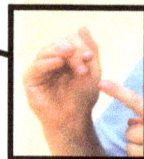

4. To complete the trick, move your holding hand AWAY from your body and keep it closed!!! (This seems to be the hardest part of this trick for people.) At the same time, make sure you POINT at your closed hand with the other hand, tricking your audience into thinking the object is still in your holding hand, when it really isn't!!

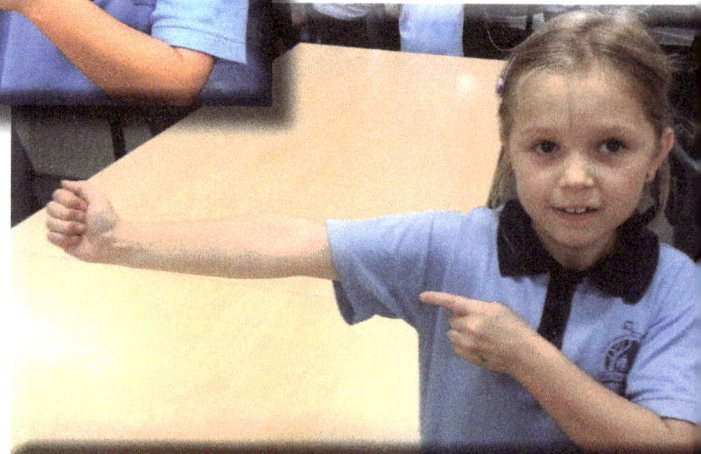

Communicate! Try and teach someone how to do this trick properly – it's not easy! But one of the best ways to learn something is to try and teach it to someone else.

Simple Spectroscopy

Every kind of atom glows in a different colour - why?

One of the oldest questions ever asked is 'What is stuff made out of?' Nowadays, we think of all objects as being made of **atoms**. Every atom, indeed every material, glows in a different colour when heated. Each glow can be broken up into its rainbow, or *spectrum*. The *science of spectroscopy* can tell us what even very small or very distant things are made of, even distant stars!

How to Make Green Fire

Only adults should do this demonstration

1. Mix a spoonful of copper sulphate (available at most garden stores) with enough water to fill up a small spray bottle.

2. Using all appropriate caution, spray the solution into a medium flame. As the fire heats up the copper, it will glow *green*!

Be Careful!

1. Copper sulphate is highly corrosive - don't spray it near other metals or they WILL RUST. Also, copper sulphate might be great for plants, but it's deadly to fish. Don't wash it down the sink.

2. We like to get a 10cm diameter metal lid or pan and place in a little methylated spirits. You don't need much, just a millimeter or

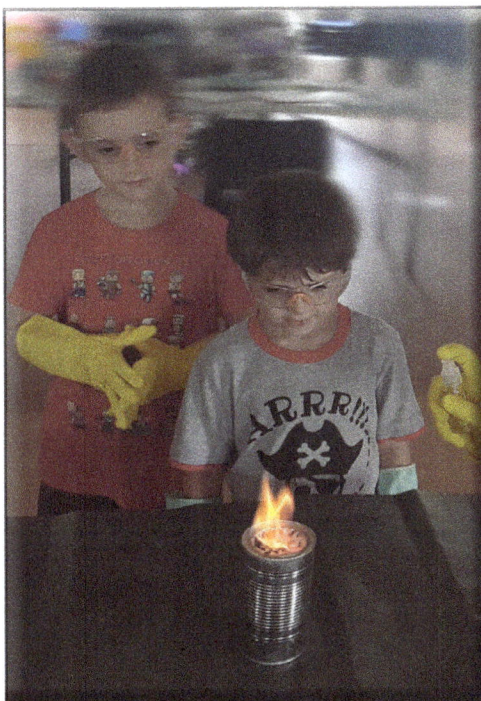

What colour is fire, and why?

two to cover the surface. Make sure you light it from a safe distance in a well ventilated (NO wind) area.

#CreatingScience
SimpleSpectroscopy

Why Does it Work?

In order to cool down, chemicals will glow in a different colour depending what chemical they are. Copper burns green, while other chemicals glow in different colours that your eyes cannot see, such as the ultra-violet glow of scalding aluminium. There are a lot of variables that affect this demo, it is just a simple start to help get you recognising atoms and the wonderful fun they can be!

What other metal-salts will work? Try potassium iodide, strontium chloride or sodium chloride (aka, table salt).

INQUIRY SCIENCE

Potassium

Strontium

Copper

Sodium

Science is People!

Every science idea, ever, belongs to a person. Who where they? What did they say? And how did they convince the rest of us that it was a good idea? Here are a few of our favourites, just from Australia:

I N Q U I R Y   S C I E N C E

Ruby Payne-Scott

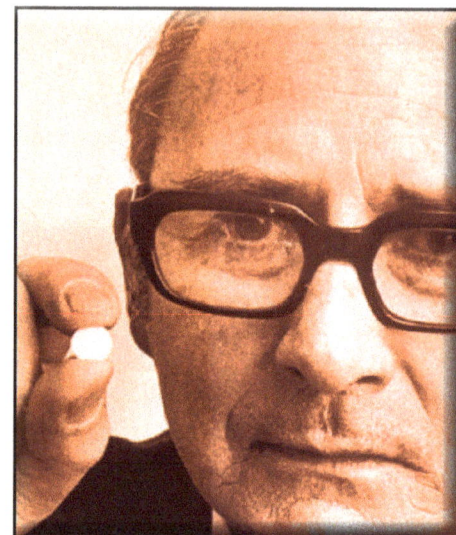

A pioneer in Radio Astronomy, and one of Australia's best and first female scientists, she is credited with the discovery of type 1 and type 3 radio bursts. She helped use radio waves to detect enemy planes in WWII, and researched the effect of magnetism on the human body.

Barry Marshal

In 1984 Barry Marshal and a handful of colleagues believed the 'heretical' notion that stomach ulcers were caused by microbes, and not stress and bad diet. He proved it (DO NOT try this at home) by drinking a Petri dish of the disease, quickly developing symptoms, and 2 weeks later eradicating them with antibiotics. Their work overturned many opinions.

John Cade

Dr John Cade, a psychiatrist, pioneered the use of medicines for treating psychological disorders in a time when the usual treatments included electrocuting people or chopping out bits of their brains! His work included the first effective medicine for bipolar disorder.

Dr Joseph Ireland

While not Australia's greatest scientist, "Dr Joe" is a social scientist who also loves to share the joy of science with school kids all over Australia.

Maybe... You?

Insert Your Picture Here

With hard work and dedication maybe the picture that belongs here will be you one day? Contributions big and small all make a difference to science!

Remember, *someone* always makes up the science - it doesn't come from nowhere! That someone will have a history, a culture, and will almost always have a question that got them started on their science journey.

Science belongs to people!

Creating Science!

Biological Sciences

Reversible Ears

Sound is an amazing thing, can you hear the world in reverse?

Build this simple toy to trick your ears into hearing things from the wrong direction! You'll need: two funnels, sticky tape, Blu Tack for safety, and two 50cm lengths of thin plastic pipe.

B I O L O G I C A L   S C I E N C E

1. Attach a funnel or roll of cardboard to one end of each piece of pipe. You may need a grownup's help. Then you should tape the funnel on securely.

2. Tape the two pieces of pipe securely together, with the funnels quite close to each other but the open ends pointing in opposite directions.

3. It makes good sense, especially if your piping is quite firm, to soften the ends with tack. This helps prevent scratching your ears, or sharing germs, by accident.

Now turn the ends around and listen. THE WORLD IS REVERSED!

How it Works:

The sound waves are collected by the funnels, and travel down the tube to your ear.

Your brain uses the slight difference in time between when the sound arrives at one ear compared to the other ear to help it determine which direction a sound is coming from. With this toy you can trick your brain into thinking the sound is coming from the opposite direction by channeling the sound into the furthest ear first. Heaps of fun!!

Tips: You may also use cardboard rolled into a funnel shape. It works best if the tube is right at the end of the funnel, not sticking out into the middle of the cone.

Ever wondered about the sound in a seashell? What is that sound? Is it the ocean? Actually, all sound is made from waves, and waves each have a certain size. That whooshing echo sound is the sea shell filtering out all sound waves that don't fit in quite right, leaving the sound waves that do, making a very strange whispering noise.

You might notice a similar effect in your reversible ears…

BE CAREFUL! DON'T shout into the cones when the reversible ears are set up. The sound is amplified by the cone shape, and can be so loud it can really HURT. Be careful!

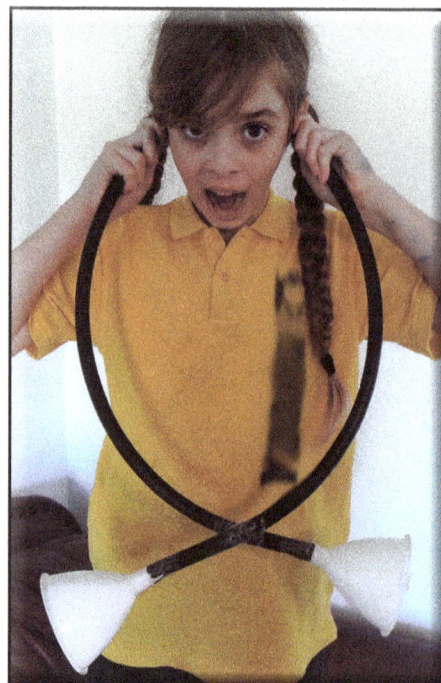

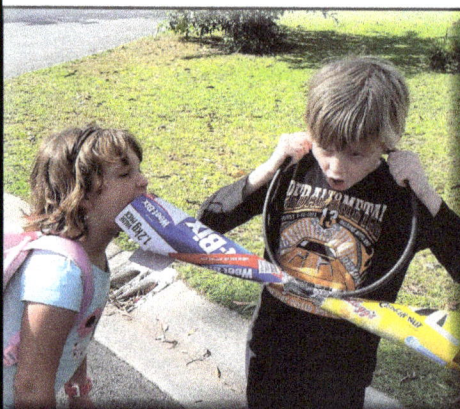

     (c) Dr Joseph Ireland 2018

Hole in your hand

What you see, and how you make sense of what you see, are very different things!

How to Make a Hole in Your Hand

- Roll a piece of paper into a tube longways.
- Place the end of the tube next to your left hand.
- Put the tube up to your right eye and look through it (not at it).
- Open both eyes! Can you see the 'hole in your hand'?

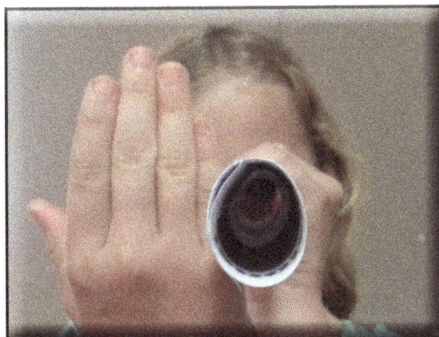

Floating Finger: Open both eyes. Place your fingertips, not touching, but pointing to each other, just in front of your face. Look through your fingers at something far away. You might see your floating, fuzzy, flying finger!

Find Your Blind Spot

Cover your left eye, stare at the circle, move your head forwards and backwards until the star disappears (at about 30 centimeters). Having trouble? Don't despair! It's very tricky the first few times! With practice you can do it anywhere, any time, easy! Each blind spot is usually covered up by our other eye, and our clever brains.

Dominant Eye

Most people automatically lift the tube up to their dominant eye. Just as we have a dominant hand, we also have an eye that we tend to do most of our looking out from. Sometimes our other eye then gets lazy and needs extra exercise to keep up!

Why do illusions work?

Our brains receive a LOT of information, all the time, from each of our dozens of senses. In order to make sense of all that information, the brain has to cut down all that data to just the important information, and along the way, it sometimes makes mistakes.

For example, you know you don't *really* have a Hole in Your Hand, but in order to process a different image from each eye, your brain can make it look like you do. Having two working eyes is very helpful for accurately judging how close a nearby object is.

BIOLOGICAL SCIENCE

Tactile Mazes

From Mechanoreceptors to Nociceptors, what is our somatosensory sense?

B I O L O G I C A L   S C I E N C E

Aristotle Illusion

Cross your fingers as shown, close your eyes, and rub a pencil in between your fingers. It feels weird! You might even get the sense of there being two pencils, not one! How??

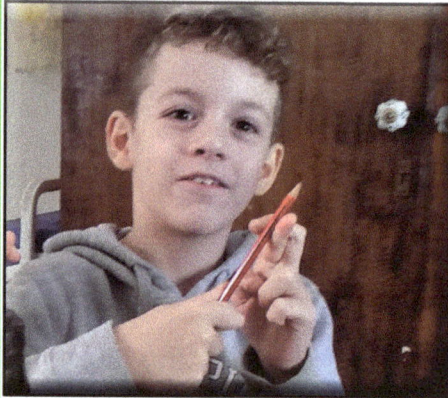

Chilly or Warm?

One well known yet very discombobulating tactile illusion is to place one hand in warm water, another in cool water, then place them both into normal-temperature water. Each hand will disagree about the temperature of the normal water!

Fooling Proprioception

Stand in a doorway and push your arms out against the door frame for at least 30 seconds – make sure you really try (but DO NOT hurt yourself!) Then step out, and your arms feel like they still want to lift up and sideways. WHY?

Tactile Mazes

Hints: Don't make them too complex. And use more than one finger to find your way - maybe even all of them, just like braille readers do.

A maze for your hands and your imagination. These are super difficult with your eyes shut! You'll need:

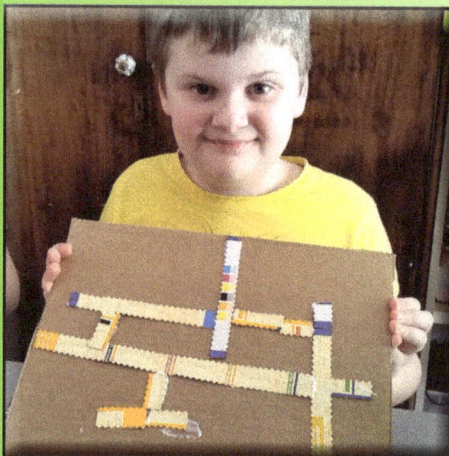

- Ridged cardboard to put them on.
- Thick cardboard to make your maze.
- Wood glue, to keep the cardboard together.
- Some tactile decorations to enhance the mazes, and to indicate a start and a finish. Velcro dots can be fun, as well as feathers, Blu Tack, or folded cardboard.

How it works: Our brain creates an image of the outside world using all your senses. When you focus on just touch it can be a little bit tricky! But with practice anyone can get very good at this game.

The part of your brain which detects touch is called the somatosensory cortex, and it includes contact, pressure, temperature, vibrations, and pain. We also have nerves in our muscles to help us know how we are positioned and moving, called proprioception.

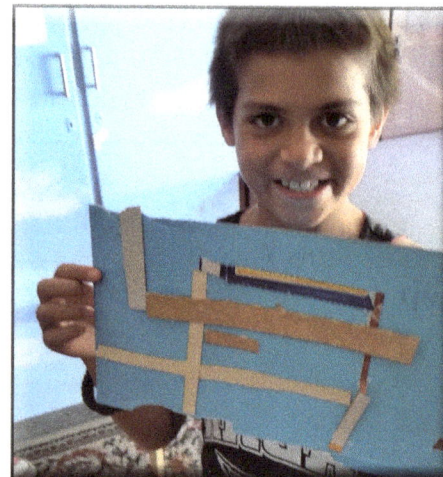

Balancing Mouse

#Creating Science Balancing Mouse

Proprioception and Balance - Senses we seem to ignore! How do we keep our balance?

Staying balanced uses two important things:

- Our middle, or rather the **centre of mass**; the mathematical point where all our body's weight is distributed evenly. When we spin in a circle, it's the part everything else spins around.

- Our **base** is the shape formed by the part of us in connection with the ground. Usually, it's a big rectangle between our feet, or the big square under the chair we're on.

Balance is simple: when your centre of mass is over your base, you are stable and won't fall over. But as soon as your centre moves away from your base, you will begin to fall!

Rolling marbles always have their centre just over their base. In a sense, they are simply experiencing guided falling.

You can shift your centre of mass by reaching a limb out, or putting something heavy in one hand. *For example*, you can lower the center of mass on the mouse on this page by putting weight on her...

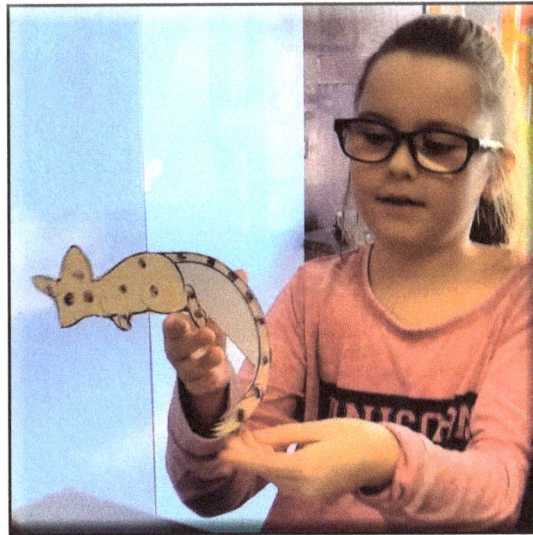

Because there's one really fascinating thing about balance - when your centre of mass is BELOW your base, you become *self stabilising*!

Staying balanced needs proprioception, balance, touch, and sight to help us know where our body is, and what it is doing. Our sense of balance is right next to our inner ear, including up/down, forwards/ backwards, and the three rotational directions.

The Hopping Mouse

Download your instructions from
www.CreatingScience.Org/activities

www.DrJoe.id.au

Catching Insects

How do we catch ants for study? Warning: Do not touch living insects

Insects are one of the most important, diverse and amazing life forms on this planet! Insects live in many different places, including desert sand and under the Antarctic ice. Without insects, just about all other life would eventually die out, including us!

This activity involves building a few simple insect traps. We will not be killing or displaying insects in this lesson; just looking at them from a safe distance and then putting them back where they came from. Research shows insects have feelings too, and so probably don't like being captured any more than you, so remember:

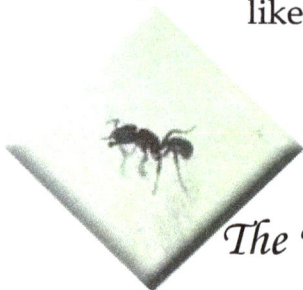

Release all insects back into their environment after ten minutes or less!

The World's Most Simple Insect Trap: The Bug Sukka!

Materials: A plastic soft drink bottle, clean and dry.
A glob of Blu Tack or plasticine
A straw, the thicker the better.

Method: Simply place the straw on the top of the bottle, making sure you form an air tight seal around the lid. You're done!

How to work the insect trap:

1. Practise squeezing the bottle, then letting it go suddenly. This creates a sudden inflow of air that can push unsuspecting insects into the bottle, allowing you to keep them for study. You might like to try holding the bottle steady with one hand near the straw, and the other hand squeezing the bottle.

2. Sneak up on a little insect. For this demonstration it's best if you look for little beetles and ants. DO NOT look for spiders of any kind, and DO NOT go turning over rocks and bark.

3. Release the bottle quickly, and with luck and some skill, you've captured a little insect. Release with gratitude after 10 minutes or so of study.

Creating Science:

Try drawing a picture of your little insect. What parts make up the body of your insect? What do those body parts do? Imagine what this insect's job is in the environment. Does it clean up rubbish, protect its hive, or simply hunt other insects? What do you think the insect is feeling? Can you make up a story of the day it was captured, studied, and released back home?

BIOLOGICAL SCIENCE

Hunting Insects:

Remember: DO NOT go lifting up any old logs or leaves, or searching through dusty and dangerous places (especially in Australia)! Stay away from places that might be dangerous. As author Douglas Adams once joked, "Nine out of ten of the world's most dangerous spiders live in Australia, though it might be more accurate to say nine out of nine of the world's most dangerous spiders live in Australia." So be careful! There are plenty of insects to be found scurrying around footpaths and up trees. Look there first.

Can't Find Insects?

Here's another simple but disgustingly effective insect trap you can use with your Bug Sukka. Get yourself an old bowl or the top of a soft drink bottle. Bury it up to the rim in the dirt, and place a few bits of fruit, honey, or some other insect treat (even balsamic vinegar). Leave overnight, and no matter where you live, often within hours) it will be crawling with creepy crawlies! Remember, don't touch the insects or the old fruit, and throw it out immediately (preferably with the help of an adult w e a r i n g gloves – this activity is sick and super gross!!!)

Always Safety first! DO NOT touch insects with your hands (unless you are a professional too), and do not go looking for insects in dark, dangerous places! Remember to let your insects go after ten minutes or they will die. Treat them with respect, and you can have fun and learn lots too!

BIOLOGICAL SCIENCE

Quadrats

I wonder what lives under the grass?

B I O L O G I C A L   S C I E N C E

Real Science

Taking a small area, and using it to estimate creature population in large areas, is a powerful way of helping scientists know about what lives in an area much too large to study any other way. Imagine trying to find every blue tongued lizard in the entire Brisbane region? Can't be done. So scientists have to estimate.

Scientists not only use this to study creatures, they use this technique to estimate plants, such as invasive weed distributions. This and similar techniques are used to help estimate the population of endangered species. Can you figure out how to use this technique to see if ant numbers are decreasing?

The process is incredibly simple:

- Mark off a small area (a 10cm by 10cm square is usually enough). This is your quadrat.
- Wearing gloves and lying on a picnic mat, begin to explore your small space.
- Try to draw a picture of every different creature you find.

Pretty simple? Now for the fun part:

- Find out how big the entire yard is.
- Over ten minutes, carefully try to number *one kind* of insect you find in your quadrat - ants, for example.
- Find out approximately how many of that kind of creature you have in the yard by multiplying the number of creatures you found, by the number of times your quadrat fits into the yard!

*It's not unusual to find there are **hundreds of thousands** of ants, beetles and bugs living in the average back yard!*

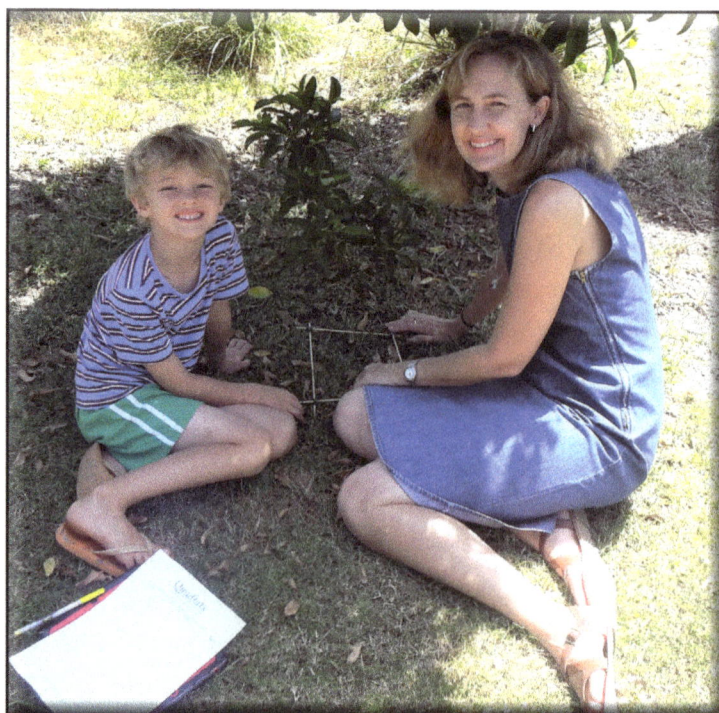

#Creating Science Terrariums

Terrariums

The simple science of life illustrated with a "biome in a bottle"! Can you grow a plant in a terrarium?

Many things grow well in Terrariums!

Suggestions:

Use gloves, and always wash dirt from hands!!

Some tough plants work great - but they can also be weeds, so use wisely! These include devil's ivy, spider fern and air plants.

You can add figurines such as dragons or fairies, and give your terrarium a fun narrative.

A little activated charcoal over the cloth will help to reduce bacteria, fungi and odors.

All you need is:

- A soft drink bottle
- Pebbles
- Sphagnum moss or old, clean towels.
- Soil
- Seeds

How to Build Your Own Terrarium

A. Start with a soft drink bottle, cut in half.

B. Put some gravel in the bottom, to show water level.

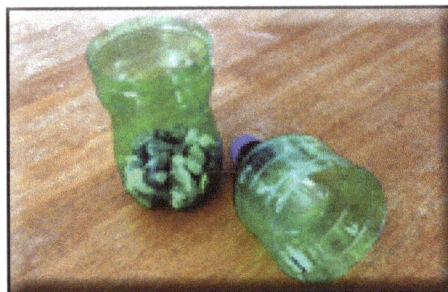

C. Place cloth or sphagnum to keep the soil out of the rocks.

D. Press down the cloth.

E. Cover it with soil, about 20cm.

F. Put in a few seeds and place the top back on.

BIOLOGICAL SCIENCE

The Needs of Living Things

What do living things need to keep them alive? #CreatingScienceNeedsOfLivingThings

BIOLOGICAL SCIENCE

Food

Water

Shelter

Books

Free Wifi

Electricity

The Sun

Needs vs Wants

Which of these things do living things need to stay alive?

Remember: We're not looking for the things that keep living things happy, only those things needed to keep them *alive!*

We're still learning about the needs of living things!

We used to think all life either got its energy from the sun, or from other forms of life that did - until we found the deep sea extremophiles! However, all living things need **food** for energy and to build and repair their bodies. They can get their food from the most surprising sources.

Living things need **somewhere to live** - a place that is just right for them. Sometimes it's deep under the sea, or in the choking blackness underground, but there's somewhere for everything! This place must have the right conditions of warmth, light, food etc.

All living things need **water** to grow and breed - at least, there's no known exception to that rule so far! But then again, science isn't finished. Even viruses need water to reproduce inside other living cells. Can you imagine a living creature that uses something other than water to drink?

Believe it or not, not all living things need air either. Anaerobic bacteria use other chemicals, and may have been the first life forms on earth.

And while it may seem unlikely to you, no living creature on earth needs Wifi to survive, but it sure can help make life fun!

Adaptation

Can life change in order to prosper in a new environment?

Imagine that long ago, little creatures called Mellits lived happily in the cool, abundant grasslands. They had eyes to see with, ears to hear with, and four legs for running around. Then one day a fierce storm blew them all over the world. Over many generations, their bodies changed to live in their new environments. How do you suppose they might have *adapted* to make sure that they prospered in their new home?

The Dry Desert

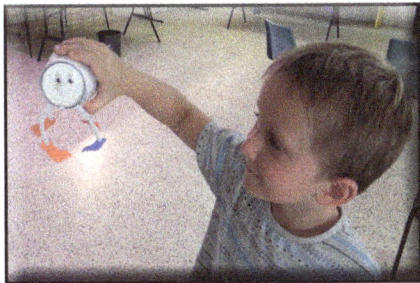

A Rare Desert Mellit

With long, long legs to keep them away from the hottest air near the sand, and shady feather coats, these Mellits will survive a hot desert just fine!

Rainy Rainforest

A Happy Rainforest Mellit

With feathers to help keep warm at night and cool in the day, this fellow also has strong claws to clutch onto slippery rainforest branches.

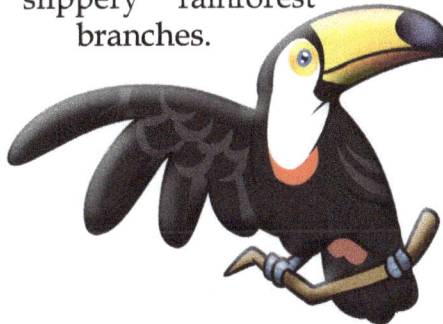

Make Your Own Mellit!

With eyes to see, ears to hear, and legs to walk, how might their species adapt to new environments? How about a swamp? Or life under your bed...?

The Freezing Arctic

A Snuggly Arctic Mellit

With much thicker coats than other Mellits, their short and stout bodies help them to keep the heat in.

B I O L O G I C A L   S C I E N C E

How it Works

Babies are always a little different from their parents. Over time, and over many generations, these little differences can really add up! Scientists understand that this is how animals can come to adapt to live in new environments, though it may take a very long time.

Trace Fossils

Does the evidence of living things long ago teach us about life today?

#Creating Science Trace Fossils

Sometimes when a creature dies its remains can be preserved. We call this becoming a fossil. We can learn about the past from them. Most animals are eaten or broken up before they can be preserved. Not one in every million living creatures becomes a fossil, and we've yet to find even more than one in every million fossils that may exist! Lets make a fossil of our own!

Trace Fossils

Why not make your own trace fossil?

Paint the palm of your hand or foot, and while it is still wet press it to a clean piece of paper. Label and date it, and after it dries you can keep it as your very own trace fossil to show you lived!

Or take some air drying clay, and press your finger into it. After it dries you can fill it with plaster to create a mould of your finger, or keep the imprint as it is. Again; evidence you were here!

What type of fossil do you think these would be called?

The fact is: most (i.e., more than 99%) of all the different species that have ever lived on earth are no longer alive today. The only way science knows that they were here is by what they left behind.

Creating Science:

How old does evidence of life need to be before it is considered a 'fossil'?

What are index fossils? How do they help us know when things lived on earth?

What proportion of fossils are discovered whole, while all others are only partial remains?

BIOLOGICAL SCIENCE

Creating Science!

Chemical Sciences

Red Cabbage Indicator

What is acidic, anti-acidic, and how can we tell?

Chemicals can often be grouped into two main categories:
Acids taste sour and are corrosive to metals.
Bases (or 'antacids') feel slippery and often taste bitter.
How can you tell what chemicals at home are Acids and Bases?

#Creating Science RedCabbage Indicator

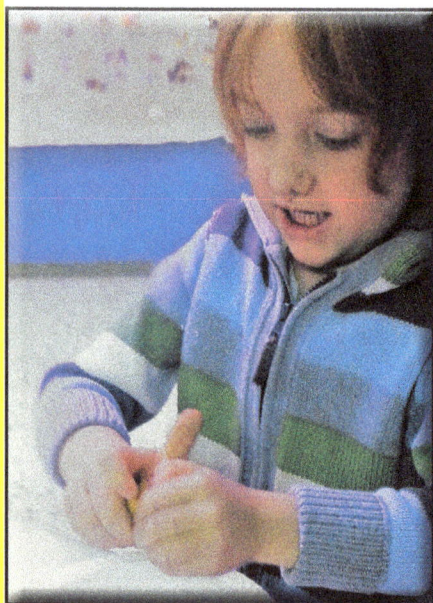

CHEMICAL SCIENCE

Sometimes squeezing lemons is hard work!

Red Cabbage Juice

1. Buy some 'red cabbages' from the supermarket. You may have to look around, but they're fairly common. You know you're looking at red cabbages because they are purple (!?)

2. Cut it into nifty little strips and freeze it. This breaks the cabbage cells and lets the chemicals out more easily. (You can boil it, but it's smelly!)

3. Mix your frozen cabbage with some water and shake well. Wait a few minutes for the cabbage juices to leak out. The water should turn purple.

4.. Mix your red cabbage juice with some household chemicals, and try to guess whether they are acids or bases. A rough guide is given to the right. Try lemon juice, vinegar, diet soft drink, tap water, distilled water, salt water, and water with a teaspoon of bicarb. Remember, below 3 and above 11 can be very dangerous: avoid them (or adults only)!!

Caution! Some acids are dangerous, and some bases are just as dangerous as acids! – Make sure you wear chemical safety gear such as goggles, plastic gloves, and a coat to protect your clothes!

Approximate Red Cabbage colours:

pH	Example	
0	Battery acid	Acids – 0 to 3 DANGER ZONE!
1	Stomach acid	
2	Lemon juice	
3	Vinegar, soft drink	
4	Tomato, acid rain	
5	Coffee, bananas	
6	Milk, urine	
7	Pure water	Neutral
8	Sea, Egg whites	Bases – 11 to 14 DANGER ZONE
9	Baking soda	
10	Milk of magnesia	
11	Ammonia	
12	Mineral lime, soap	
13	Bleach, oven cleaner	
14	NaOH, drain cleaner	

#Creating Science Sherbet

Sherbet

What chemical makes the FIZZ of sherbet?

Taste and Smell are, in many ways, exactly the same sense - **Chemoreception**

Materials:

- Citric acid
- Bicarbonate soda
- Icing sugar
- Flavoured jelly crystals
- A teaspoon
- A zip lock bag for mixing and keeping your sherbet
- Paddle pop stick for eating sherbet

1. Add 1 flat teaspoon of citric acid crystals to the plastic bag.

2. Add 1 flat teaspoon of bicarbonate of soda to the plastic bag.

3. Now add about 3 heaped spoons of icing sugar. MIX THOROUGHLY!!

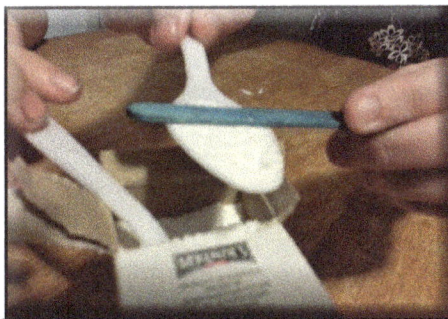

Advice:

1. Eat sherbet in small doses! Too much and when you close your mouth it might turn into a dust that flows easily down your windpipe and you might start choking!

2. Have some water handy. Not only to wash down sherbet dust, but also because if your sherbet is too tart, or you didn't mix the bicarb properly, you'll be grateful to have something to wash your mouth with!

Why it Works:

When you combine an acid (in this case the citric acid) and an alkaline (the bicarbonate of soda) with water (in your saliva) they mix together to create a gas called carbon dioxide. You are actually feeling the sensation of carbon dioxide bubbles on your tongue.

This is the chemical that makes fizzy drinks fizzy!

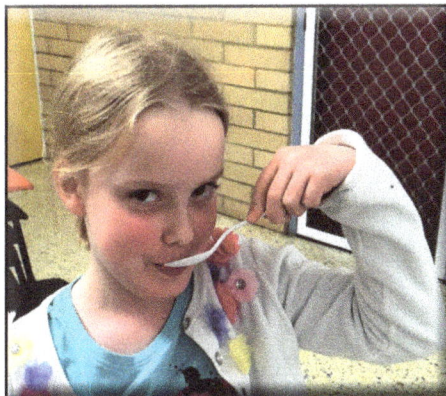

Creating Science

1/ What effect does adding a heaped spoon of jelly crystals have?

2/ Can you find out how sherbet was first invented?

3/ What other uses are there for carbon dioxide gas? Is it helpful, or harmful?

CHEMICAL SCIENCE

#CreatingSciencePlastics

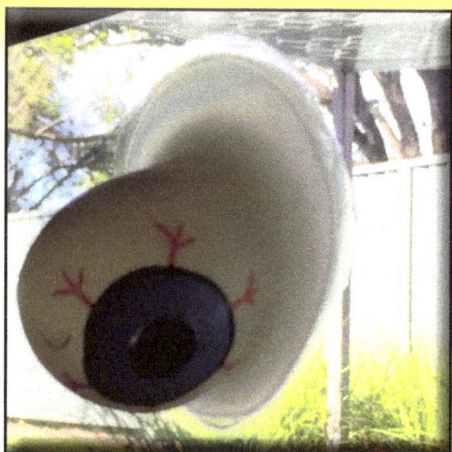

Plastics

Cool Plastic Questions:

(Find the best answer!)

1/ Where is the largest plastic rubbish tip?

2/ Who invented plastic?

3/ What does the work 'plastic' mean?

4/ How long does plastic take to degrade?

5/ Are there 'biodegradable' plastics?

C H E M I C A L   S C I E N C E

A/ It comes from the Greek πλαστικός (plastikos) meaning "capable of being shaped or moulded". Not all plastics can be reshaped by heat.

B/ It is in the north Pacific ocean, and when you fly over it you probably can't see any plastic as it is almost all microscopic pieces.

C/ Parkesine is generally considered the first man-made plastic patented by Alexander Parkes in 1862.

D/ It depends on the kind of plastic, but it can be around 1000 years! And it usually doesn't biodegrade, it just photodegrades into smaller pieces.

E/ Yup, made from organic and inorganic materials, these plastics will break down into carbon dioxide, water, and other natural inorganic substances, enough given time.

Did you know scientists are trying to invent plastic eating microbes!

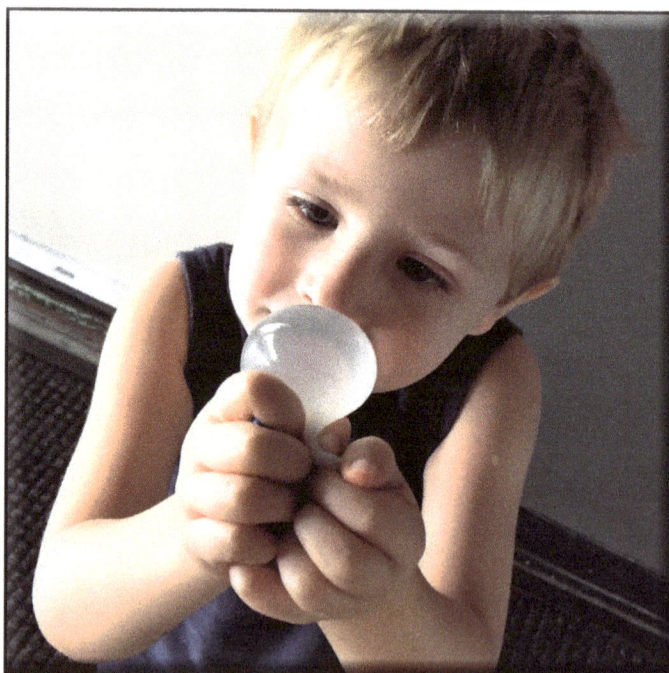

Know Your Plastics

What does the number mean? NOT the number of times it can be used before it should be thrown out (though that does make an approximate guide occasionally)

#1 Plastics: Polyethylene Terephthalate (PET or PETE). Commonly considered safe.

#2 Plastics: High Density Polyethylene (HDPE) HDPE is a sturdy and reliable milky plastic. Dishwasher-safe and able to stand temperatures from -100 to 80° C.

#3 Plastics: Polyvinyl Chloride (PVC). Products made from PVC should not be used to store food.

#4 Plastics: Low Density Polyethylene (LDPE) Resistant to breakdown due to chemicals from acid, oils, greases, and more.

#5 Plastics: Polypropylene (PP). Its durability makes polypropylene plastic a good option for reusable bags and food and drink storage.

#6 Plastic: Polystyrene (PS) Polystyrene is both potentially toxic and widely overused. Avoid!

#7 Plastic: Everything else, including Tupperware. These containers can be any of the several different types of plastic polymers.

Answers: 1=B, 2=C, 3=A, 4=D, 5=E

Gak

A very fun, and very special kind of goo. In 1943 James Wright, an engineer, was attempting to create a synthetic rubber. He failed, and kept the Gak on the shelf. It wasn't until someone else paid him a visit, and had a play with the goo and thought, 'Hey, kids would love this!' that the goo was turned into a science toy. Now it has seen use as a grip strengthener, as an art medium, and it even went into space on the Apollo 8 mission! Can you make the simple plastic, and investigate the properties, of Gak?

1. Get some Borax (the cleaning powder), sealable plastic bags, and some PVA glue (polyvinyl acetate, or craft glue).

2. Place about ¼ of a teaspoon of Borax in the bag. Make sure you clean the spoon right away!

3. Add about ¼ of a cup of water. Mix it together to get the water to dissolve most of the borax.

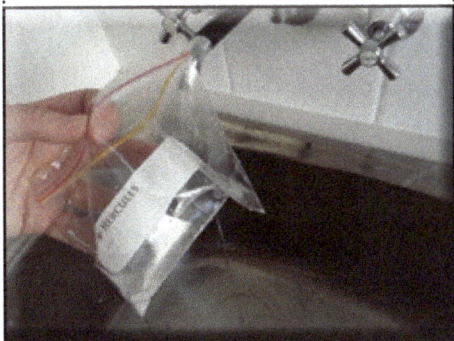

4. Next add about ¼ of a cup of PVA glue.

5. Mix thoroughly. All the glue must react with the borax. If it's gluey, add more borax.

6. Squeeze out any excess glue (and borax) THOROUGHLY under running water. Wash your hands.

Congrats, you got Gak!

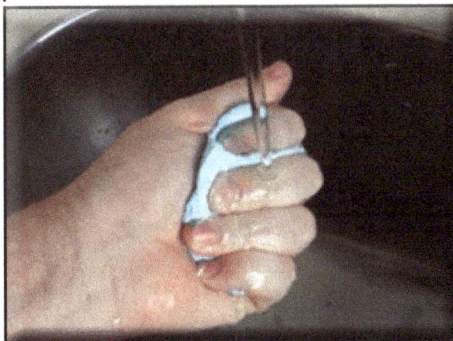

Experiment with your gak - make sure you take notes! What happens when you:

Bounce it, squish it, pull it, put it in water overnight, roll it in a ball, rest it on a newspaper, and on and on!

Gak Worthy Notes:

- Keep it clean! Oil from your fingers, and water from the environment, can encourage mold to grow. You can wash your Gak with soapy warm water any time.
- Keep it in a plastic bag or cup! Gak loves to soak into things, like clothing or furniture. Don't let it!
- Nothing lasts forever, not even Gak - but it *can* last for centuries! So when you're done with it, make sure you put it in the rubbish bin!

CHEMICAL SCIENCE

How it Works: Very Much like Wet Spaghetti!!

The glue forms long, long chains of molecules called polymers. *These are like strands of spaghetti. The borax forms millions of very weak links between the long strands. This is like the spaghetti water. When spaghetti is hot and wet, the strands all slip and slide over each other, just like in a liquid. This is also what happens when gak is flowing; the polymers slide over each other and there aren't too many bonds to hold them. When the bonds have time or enough pressure to form many, many links, it holds the spaghetti together like a fat, bouncy brick. Not that I'd recommend it, but did you know that a big ball of damp, cold spaghetti left out to partially dry can actually bounce?*

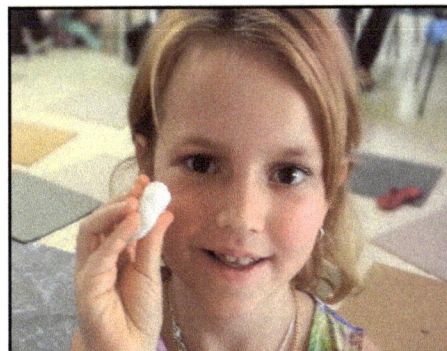

Magnets

One of the most amazing and useful tools we have today, and even a thousand years ago, is a good, reliable magnet. But what are magnets? While they are old, there are still a lot of mysteries surrounding them, not to mention a plethora of misunderstandings! Are all metals magnetic, and can we make something magnetic?

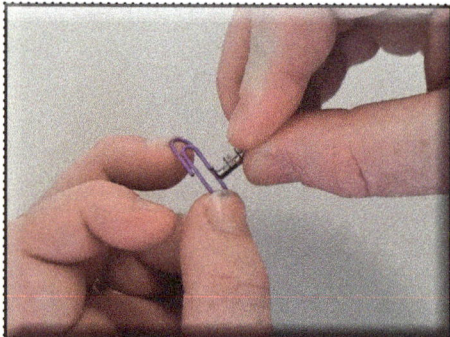

C H E M I C A L   S C I E N C E

1. Stroke your paperclip with the magnet at least 30 times, in the same direction each time, NOT back and forth!

2. Rest your paperclip in an upside down lid - make sure the lid doesn't stick to a magnet or it wont work!

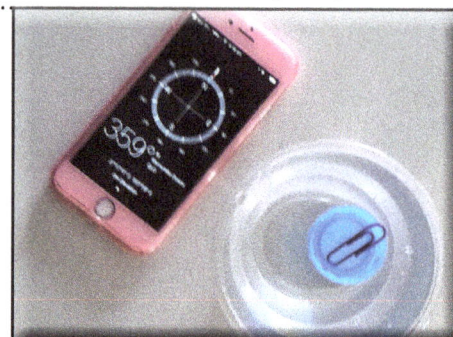

3. Overflow a cup with water, then float the lid on a full cup and you've got yourself a super simple compass!

A thousand years ago, this device you can now potentially whip up in under five minutes was worth quite a bit of money and expense! Magnetism is thought to be caused as the electrically charged particles in a substance line up just right. Do you know what the largest magnet on earth is? It's the earth itself. The rotation of our massive liquid nickel/ iron core creates a huge magnetic field that reaches far out into space – helping protect us from some of the dangerous particles of the sun and causing the bright lights we call the Aurora Borealis and Aurora Australis.

Can you name the current strongest kind of permanent magnet?

Neodymium = $Nd_2Fe_{14}B$

What were the first magnets made from?

Rocks. To be specific, magnetite, a magnetised form of hematite (a black rock with lots of iron).

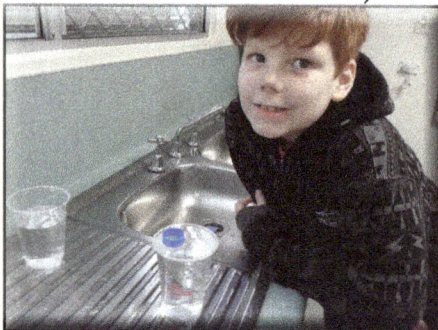

Fill the cup up to the top!

Magnetic Questions!

Are all metals magnetic?

Actually, there turns out to be only one kind of strongly magnetic pure metal, and that's *iron.*

Are magnets the only kind of magnetism?

No way! There's ferromagnetism, due to iron. There's electro-magnetism as electricity flows through wires. Diamagnetism is a property which repels magnetic fields and can be used to look into the human body in MRI scans, or to float such things as super magnets or mice! And more!

Does banging a magnet cause it to lose its magnetism?

Yup – eventually!
The material is still magnetic, but the magnetism is all mixed up and won't work properly.

Can you figure out how to get the paperclip to float without the lid? Carefully place the paperclip on some tissue paper, lower it in, gently push the tissue into the water and the paperclip will still float! Thanks to surface tension (now, what's that?!)

 (c) Dr Joseph Ireland 2018

#Creating Science Electroplating

Electroplating

It's used in medicine, makeup, and even on race cars! Electroplating – attaching one substance to another substance using electricity – has millions of modern uses! How can we electroplate a coin?

(WARNING: This is not a simple activity, and copper sulphate can be dangerous. Use gloves, proper caution, and adult help at ALL times. Dispose of the solution onto plants, *not waterways*). Material list online.

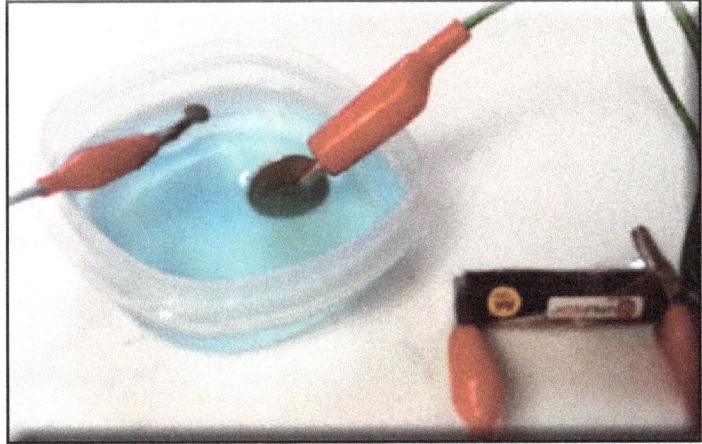

Electroplating:

Clean a metal object (coin or key) thoroughly with soap and water and then clean it with metal cleaner (but NOT metal polisher, any idea why?). Make sure you dry thoroughly.

Tape the negative terminal of the battery to the object (in our case, a coin) using a metal wire.

Make sure your anode (a nail) is firmly taped to the positive terminal of the battery.

Lower your metal object into a solution of a little copper sulphate and water. The closer the coin is to the anode, the quicker the effect will be. HOWEVER, if the coin is too close, the coin will quickly be covered in a dark black sludge. Patience is called for, not heaps of electricity!

After about two minutes you should have a nice, thin coating of copper on your metal object. You can polish it, but be careful! The coating is very thin, and it's not hard to polish the coating right off. This activity is more for illustration purposes than lasting presentation.

Handle everything with rubber gloves, and wash everything before and after.

Setting up Electroplating

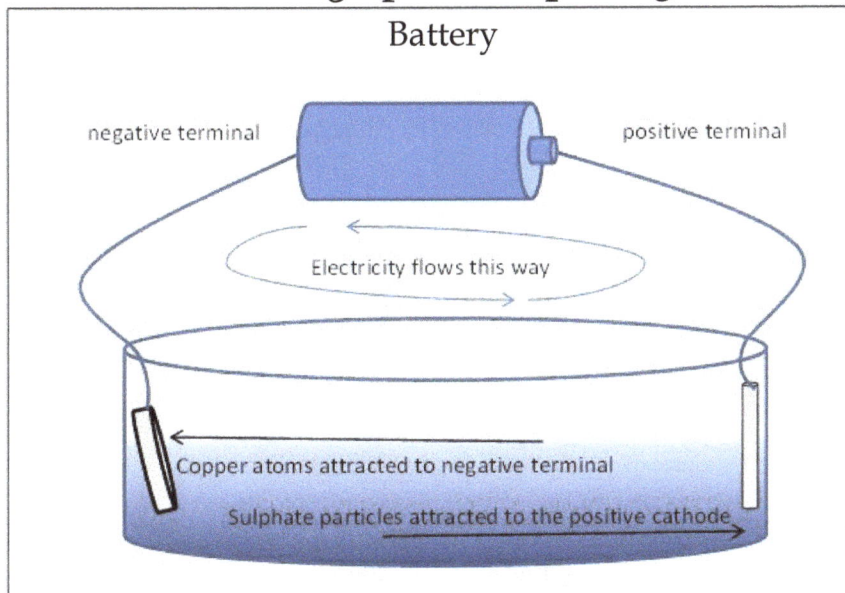

How it Works

Simple! The blue solution has copper in it. The electricity flows out of the battery, pushes the copper over to the coin and sticks it there on its way back into the battery.

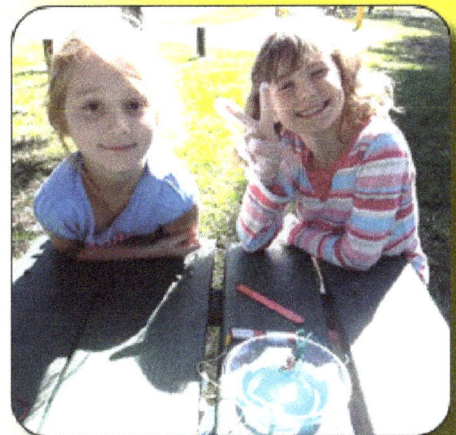

CHEMICAL SCIENCE

Bubble Worm

How do we make a bubble worm?

CHEMICAL SCIENCE

Bubble Worm Maker!

1/Take a bottle (or a cup will do) and put a little hole in one end, big enough to breathe though.

2/Cover the other end with an old towel and hold it on there - an elastic band should do the trick just fine!

3/ Blow in the end and make yourself your very own giant bubble worm!

Did you know?

Bubbles are made up of three layers - a layer of water sandwiched between two layers of suds!

Make a Bubble in a Bubble!

Anything wet with suds can go into, and out of, another bubble. Can you blow a bubble inside another bubble by making one big bubble, and using a soapy straw put another one inside the first?

Remember!

Do not share bubble makers! You don't want to mix up germs from your lips to other people.

Make a bubble bridge by putting a large bubble between two sudsy mountains!

Practise blowing up your bubble worm - remember big breaths!

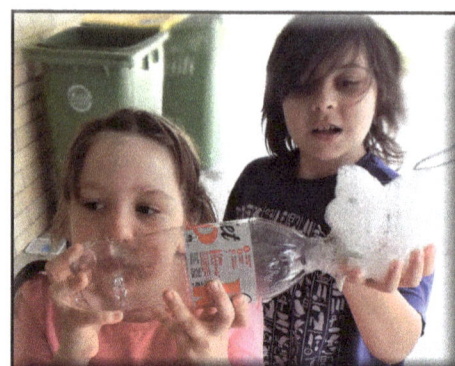

#Creating Science Density Cylinders

Density

Why do things float and sink? Can water float?

Density Cylinders - Beautiful, but messy! Take care, and dispose of down the sink!

Preparation

- Water, lots of it.
- Towels for cleaning up spills – many towels.
- Small plastic cups.
- 1 teaspoon for each group.
- Rinse cups, to rinse out the teaspoons.
- A long, thin, container. A narrow flower vase will sometimes do the trick.
- A 'float' – polystyrene balls, matchsticks, or floating beads, ~10cm.

Activity

- Find out how much water you need to fill up your container. Divide it into a handful of cups.
- Make each cup of water a different colour.

At the moment each liquid is about the same **density**. You need to increase the density – and a simple way to increase water's weight (and not its volume), is to add salt.

Salt will hide among the water particles, fitting quite neatly. This makes the water heavier with the extra salt, but its size does not increase much at all !

- So beginning at the colour you want at the top, add NO SALT. Moving down the line, add one spoon of salt, so: 0, 1,2,3,4,5. MAKE SURE your spoons of salt are all exactly the same size, accuracy is a virtue in science!
- Now place your *float* in the bottom of the container.
- In order, *slowly* pour each colour in, so from 5 to 0.

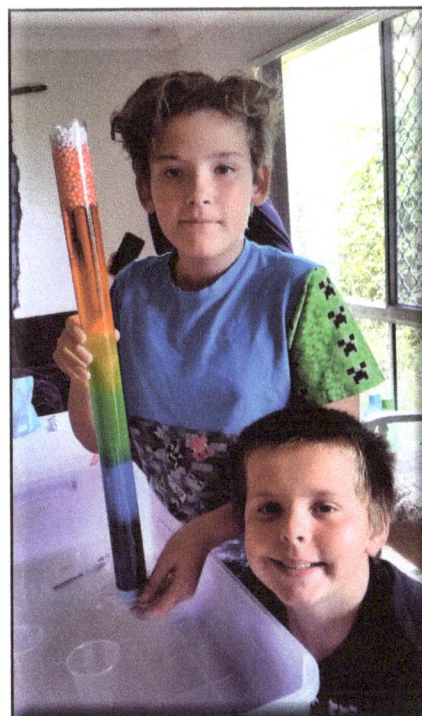

It is necessary to stop the different densities (and thus colours) from mixing together and becoming brown. ***Pour slowly!!!***

If you are careful and accurate, you should have your own colourful density cylinder soon!

P.S.: Salty water down the **sink only** *please!*

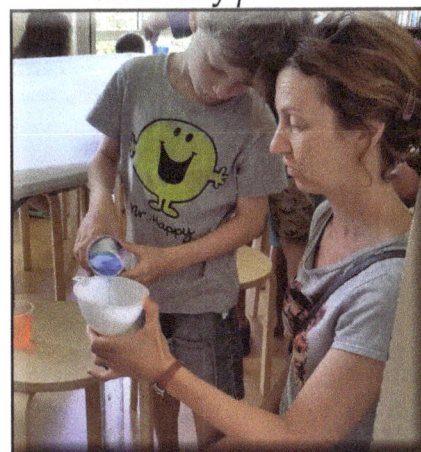

CHEMICAL SCIENCE

#Creating Science Measuring Viscosity

Viscosity

Can you determine the speed the marble moved through the liquid? Divide the distance the marble moved (in meters) by the time it took (in seconds) to get its scientific speed.

CHEMICAL SCIENCE

Viscosity = "Thickness"

Question: Drop a marble to find out which is the *thickest liquid*.

Hang on! Before you get started let's think about what we're trying to do. Scientists aren't happy with just throwing marbles around, we want more precise answers! Are you willing to help?

Try some honey, milk, oil, or even Oobleck (1 part cornflour and 1/3prt water)

Oobleck - the chemical with variable viscosity! Just add water to cornflour...

Fair testing

When we do science we want to make sure we're testing what we set out to test. What if you wanted to find out which car was fastest, but tied bricks to one of them? Not fair! So what about our marbles?

- Should we use marbles of different sizes or weights?

Keep everything the same except the thing you want to measure.

Exact measurements

Scientists like their knowledge to be the best it can be – the most thorough, accurate and indisputable knowledge possible! One way to achieve this is through embracing the virtue of exact measurements:

- Make sure you fill the container with the exact same height of liquid each time.
- Make sure to drop the marbles from the exact same height above the liquid each time.
- Measure the time it takes the marble to hit the bottom of the cup to the hundredth of a second.

Repeated measurements.

Is once really enough to be sure of your answers?

Let's embrace the virtue of repeated measures! That means testing things as many times as sensible before making a conclusion. So how many times is sensible? That's a tricky question that depends very much on what we're trying to test, but the answer is always 'the more the better!'

- Perform the activity multiple times, and take the average result.

NOW you're ready to create some great scientific knowledge!

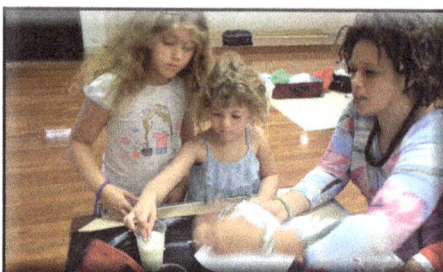

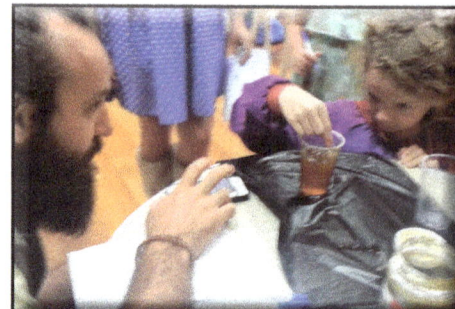

#Creating Science PuffCup

Puff Cup

Turn an old cup and empty lunch bag into a cool science toy! Can you feel air pressure?

1. Tape the edge of the bag on to the edge of the cup. Make sure there's free bag!

2. Be thorough! Lots of tape - don't let any air escape!

(3. Optional extra science; have a grownup drill a hole in the bottom of the cup).

How to Work it:

1/ Push the plastic into the cup, yet it resists… why?
2/ Now pull the plastic out of the cup, again it resists … why?
(Remember – air never sucks! It only ever pushes!)

How it Works

Air has pressure: every single particle of air is moving around, and when a particle hits something, it gives it a little push. Now even though those little pushes are so teeny tiny that you cannot see, hear or feel them individually, there are actually so many particles of air that all together they can give even a very small thing a very large push!

1. When you are trying to push the plastic into the cup, the air inside the cup now has more pressure than the air outside the cup – that is, because you're squeezing down on them the particles inside the cup tend to press against the plastic just a little more often than the air outside the cup. You experience this as a push. So the air inside the cup is pushing out, making it hard to get the plastic in. That's not so hard, eh?

2. So why is it hard to pull the plastic back out again, especially since air never sucks? One explanation is that as you try to pull the plastic out, the air in the room is now pushing against the plastic harder than the air inside the cup. This means you're trying to pull not only the tiny, insignificant weight of the plastic, but the enormous, significant pressure of the air touching the plastic as well. And as we know, air is always pushing in all directions all the time … very powerfully!

Yet when you uncover the hole in the bottom of the cup, the trick no longer works! Can you explain why?

Can you use this little hole to trick your friends?

CHEMICAL SCIENCE

Creating Science!

Physical Sciences

Making Rainbows

Can you make new colours, or see the rainbow inside every light?

How to Make a Colour Spinner:

1. Glue a circle of paper onto thick cardboard, and colour it with three even colours using markers (such as, red, yellow and blue).

2. Pierce the cardboard with two holes near the centre, at about 1cm apart. Thread about a metre of string through it and tie it off into a circle.

Now for the tricky part:

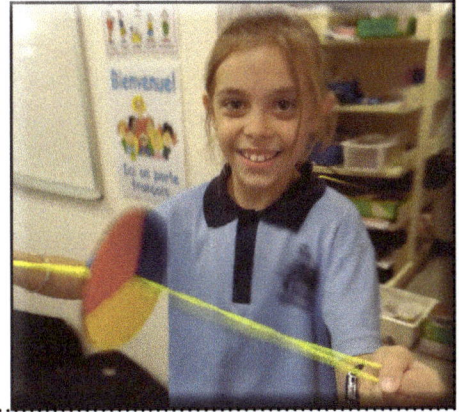

3. Spin the wheel around a few times. You can get the wheel spinning by itself by pulling then relaxing the string rhythmically. Practise!

Make a Rainbow – Sun and Hose Style

1. Make sure the sun is behind you.
2. Squirt a spray of water out in front of you, usually just off to the side a bit
3. And hey presto, you will see a rainbow!

Again, each tiny little drop bends the light. Different colours will bend different amounts, and because there are so many drops sharing the light in so many directions, you will see a rainbow. And believe it or not, all those colours are in the sunlight all the time!

Holographic Diffraction Grating Glasses

These glasses send different colours out at very different angles, allowing you to see the rainbows that are in every light. Note that sunlight has a very different rainbow compared to artificial lights - while smartphones and TVs are actually only ever red, green and blue!

Follow the rainbow around, what do you see? Rainbows are actually circles! (The horizon just cuts off the bottom half.)

PHYSICAL SCIENCE

Jumping Fun Frogs

What makes things jump?

P H Y S I C A L   S C I E N C E

This is a fun science activity that can be used to illustrate force and motion, energy, and the science of jumping!

Building the Fun Frog!!
Materials

- Jumping fun frog picture or similar (teacher notes)
- Scissors, sticky tape
- Colouring pens
- Thin rubber bands

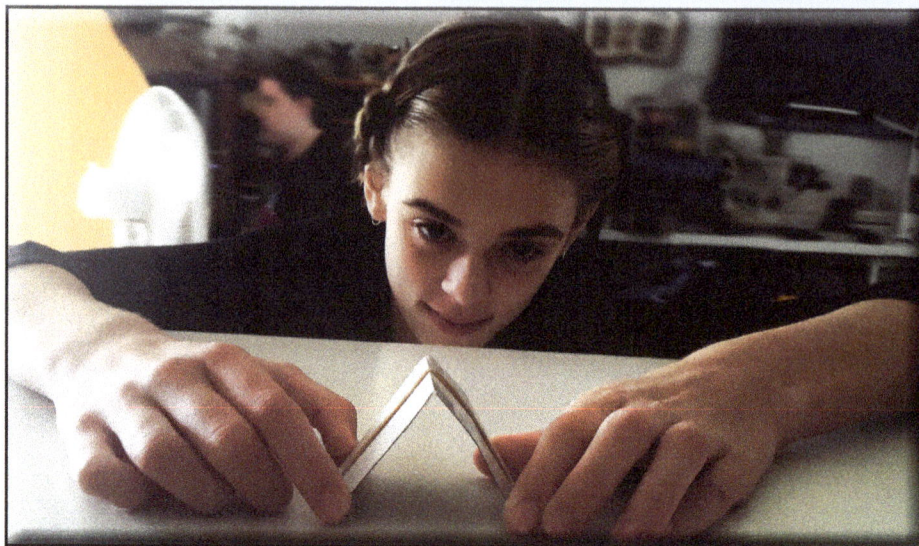

Building the Frogs

*Note: A push or a pull is **force** that makes things move!*

1. Copy or paste the frog pictures onto some rigid cardboard. This is the first big challenge of the activity – the cardboard needs to be fairly rigid or the frogs will not work properly. On the other hand, it cannot be too heavy or the frogs will not jump very high. Try experimenting to see what works best for you.

2. Stick one end of the rubber band to one side of the frog. Stick the other side of the band to the other inside of the frog, so that it sits like a capital A from the side. You can try the stick right through method or the loads of sticky tape method.

You're Ready to Jump!

Push down your jumper so the rubber band is stretched to the limit, then let go suddenly - sproing!!

Extension: This is a good chance to explore **variables** – what makes the frog jump higher, and how will you measure and record your results? What differences do the following make:

- Tension of the rubber band.
- Does stiff cardboard work better than flexible material?
- Is a big rubber band better than a small one?
- Is a loose rubber band better than a rigid one?
- Can you decorate Mr Frog and is he more 'fun' if you do?
- Etc. Ad infinitum! (Let us know if you think of a different one!)

**#Creating
Science
Lever
Catapult**

Lever Catapult

Now, let's study levers. Can you make the world's most simple catapult?

SAFETY FIRST: Stand well back from the launch zone, and use safe, soft projectiles such as tissue paper and ping pong balls! Be very careful where your projectiles will land: is the area clear of people and other things that might break, such as people or glass windows? - Safety glasses recommended!

You'll need a cup cake holder, some glue, a ping pong ball, and a ruler.

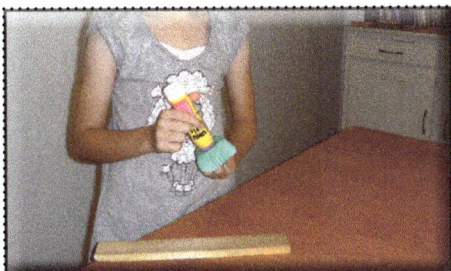

1. Choose a safe, quiet place for this activity. Glue the cup cake holder onto the ruler.

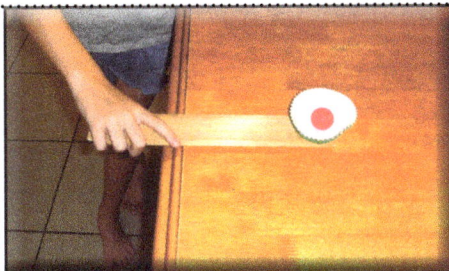

2. Place the ruler on the table as illustrated, and load it with the ping pong ball.

3. STAND WELL BACK and launch your catapult. Can you land exactly 1m away?

DO NOT HIT THE CATAPULT WITH ALL YOUR FORCE!!!

We are aiming for accuracy, not power, in this activity

Knowing how to get a projectile to travel properly is one of the most important technological problems in history – especially sport and war! What things can you try to change to make your ping pong ball travel the most accurately? Remember that we call those 'things' in science "variables".

Launch angle?

Airfoils and streamers?

Projectile size?

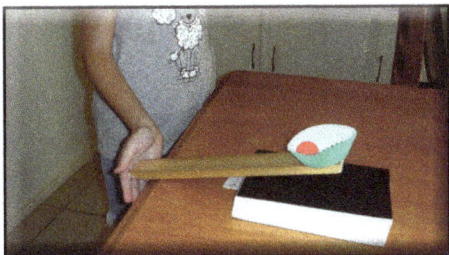

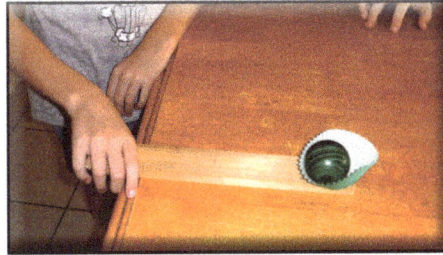

Forces makes things move, but Isaac Newton made up some more helpful rules than just that!

1/ Inertia.

2/ Forces

3/ Reaction force

Things will just float along unless a force changes that.

A force is equal to mass times acceleration.

Forces are always in pairs. Always.

Things don't stop, or start, or change direction just because they *want* to. They always do so because something has *forced* them to. Otherwise they just continue to float along. Are you on a moving planet? When will it stop? Never - until something forces the world to stop - thankfully that's not likely to happen any time soon!

So a heavy object is harder to stop than a light one, just like a fast object is harder to stop than a slow one. Likewise, changing an object's speed or direction A LOT takes more force than doing so only a little. So the force an object has is equal to how much it weighs (in kilograms), times how much it can speed things up (in meters per second squared).

So every time there is a change in direction or speed, you can be sure something else felt that same force too, but in the equal and opposite direction. For example, to catch a ball you have to push it, but you feel it push back against your hand. And to push a rocket up, you have to push something else down - usually glowing hot steam!

PHYSICAL SCIENCE

Hovercraft

Ever wondered how hovercraft work? It's pretty simple science, floating along on a cushion of high pressure air. Like to know how to make your own? Learn one way to reduce friction.

P H Y S I C A L   S C I E N C E

Warning

This activity uses balloons and hot melt glue guns – please exercise all appropriate caution.

Making the Hovercraft

Glue an old pop-top lid (or a normal lid with a hole drilled in it) onto an old CD.

Place a balloon onto the lid - you're good to go!

What Makes It Work?

Air rushing out of the balloon creates a region of high pressure between the table and CD, helping to lift the 'hovercraft' up just a little. This lowers friction with the table and allows it to slide around quite easily!

Can You Make It Work Even Better?

1/ Keep it smooth - Uneven contact between the table and CD will allow more air to escape from one side than the other, making the lid drag along unevenly.

2/ Weight is a big factor, for instance, a bigger balloon can carry more air, and more air pressure, but will it be too heavy to work at all?

3/ A second influence is the size of the hole letting the air out of the balloon. Too small and not enough pressure is built up. Too large and it will fall over every time, maybe even fly around the room! Try out different sizes and see which works best for your hovercraft.

4/ Can you think of any other variables which make the hovercraft work better?

What do you think of the idea that an object in motion will remain *forever* moving in a straight line... unless something pushes or pulls it?

This idea is called INERTIA.

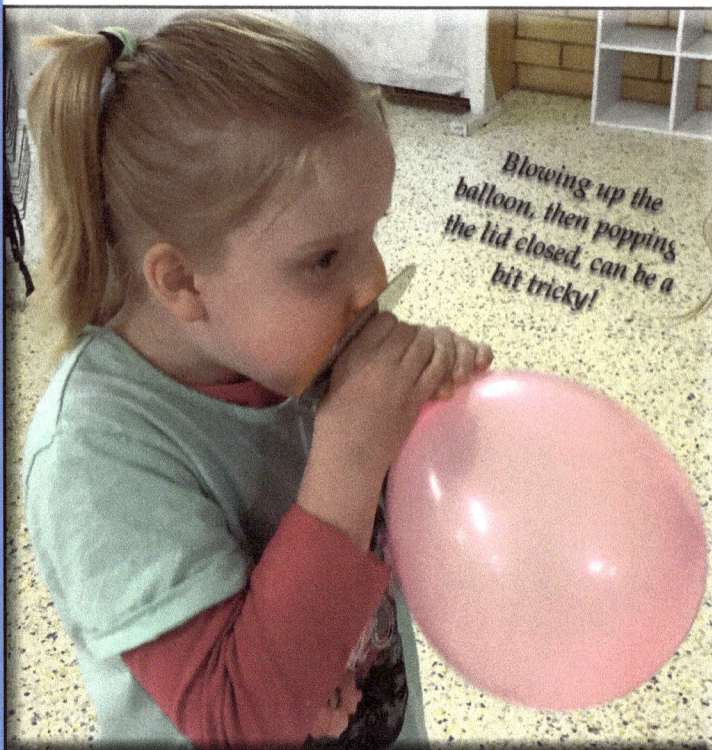

Blowing up the balloon, then popping the lid closed, can be a bit tricky!

Friction Climber

Can friction actually be useful? It stops us moving, but can it also be used to help move things?

How To Make It:

1. Cut out a picture of a super hero, cartoon character, or even a photo of a relative! This will be your friction climber.

2. Glue your picture of a friction climber onto a stiff piece of cardboard - not too thick or it'll be too heavy, or too thin and it'll be too bendy!

3. Cut two small pieces of straw into thin tubes about 2 cm long, about this long: ——————— and then tape those straws down onto the back of your friction climber at a 45° angle, as below.

4. Get a piece of string about 2 or 3 meters long and thread it through both pieces of straw, then tie it into a big circle.

5. Wrap the end of your string around a door knob or something similar.

6. Make sure the string is gently pulled tight so that the strings are now straight, and pull each hand backwards and forwards alternately, as if you're milking a cow. Use long, constant strokes and keep the string tight, but not too tight! Your friction climber will wiggle right on up!!!

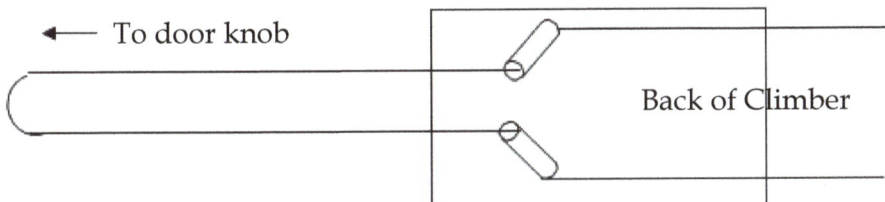

← To door knob

Back of Climber

Do You Know Why it Climbs?

Friction is a force that resists motion, and turns it into heat. This means friction pushes against any force it experiences. As you pull one string, it turns the climber around till the string is pointing straight out one of the straws and not rubbing against (almost) anything. This means there's nearly no friction in one straw, and the string there is free to move.

However, the other string is now greatly turned, and there's lots of friction preventing the string from moving through its straw. So instead of sliding through the straw (which is what it would do if there was no friction), the string now pulls the whole climber up along with it - with a little help from friction!

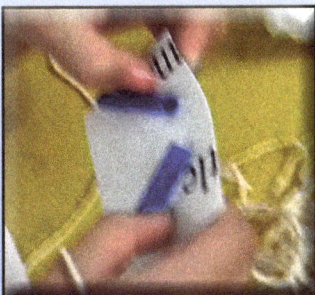

#CreatingScienceFrictionClimber

PHYSICAL SCIENCE

Gravity Falls

Why do things fall down? Is "we don't know yet" an acceptable answer?

P H Y S I C A L   S C I E N C E

1. Get yourself some old, clean, materials such as bottles, cardboard rolls and packaging to make your *gravity falls*. Lots of sticky tape is helpful as well!

2. A big piece of cardboard is helpful to stick your gravity fall on - and make sure your rollers such as marbles will fit! It might help to start on the floor.

3. With the help of some friends stick it all together - make sure you have a slope so that the rollers can be affected by the gravity of your *gravity fall!*

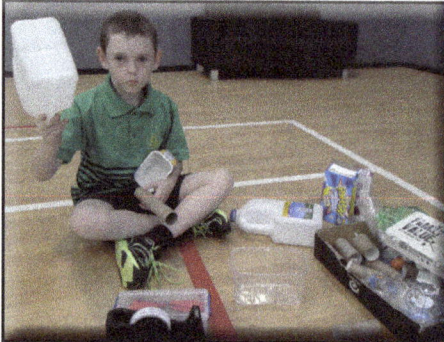

4. With practice, your rollers can be doing all sorts of tricks!

Remember – gravity doesn't only "pull things down", that's just a side effect. We should probably say gravity "pulls things together." Remember, all forces are in pairs! If the Earth pulls your body down, then your body is also pulling the Earth up, but since the Earth is so much larger it 'wins' every time!!!

Gravity Questions!

What falls faster - light or heavy objects? Well, ignoring the effect of wind resistance, Galileo from Italy taught that they both fall at the same rate! *Incredible*!

Will your marble run work on the moon? Of course it will! There's gravity on the moon, just less of it than on earth - so it will run, but slower.

Is there gravity in space? There sure is! Gravity is what holds the earth in orbit and the galaxy in its shape. We just don't feel gravity when we're in orbit.

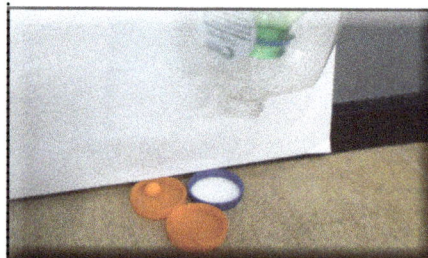

So what *is* gravity?

Insulated Houses

What is heat, how does it travel, and how do we stop it from moving?

Long ago, great scientists believed that heat was probably an invisible goo whose presence made hot things hot, and without it things felt cold. They called it "caloric". Not every scientist was happy with this idea. For example, why did drilling a cannon, taking away material, make it hotter and not cooler? It took hundreds of years to convince some scientists that heat is more accurately described as how much the invisible particles are jostling about. We call this idea the **kinetic model of heat**.

Which shoebox warms up fastest? Which shoebox cools down quickest?

Black or White? Which is right?
Both have an important role to play in keeping us insulated!!

Why is This Balloon Immune to Fire?

If you look carefully, you can see a puddle of water in the balloon. As the fire heats up the balloon, the water carries away most of this heat, preventing this balloon from melting and then exploding.

How Do We STOP Heat From Moving From One Place to Another?

1. Have thick walls - to slow the conduction of heat from direct contact. Thin walls will cool you down or heat you up, sometimes very quickly! Things like fridges and eskies all have thick walls to slow **conduction**.

2. Stop **convection**. Double glazed windows can help the hot glass from heating up the cool air inside the room, or vice versa, because they are filled with almost nothing in between the panes of glass. Having no windows could be better, but what fun is that!?

3. Use reflective materials to stop the infrared **radiation** from heating things up. And if they do get hot, use colours that cool down fast - black is best!

Insulation is very important!

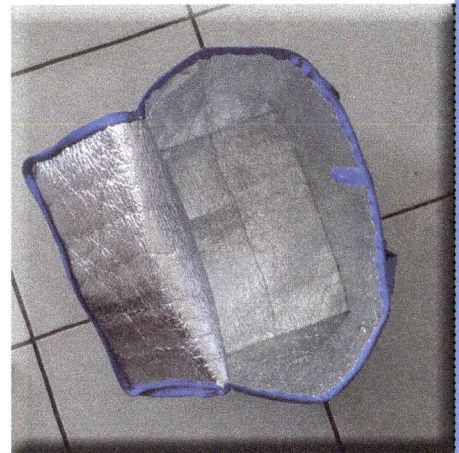

PHYSICAL SCIENCE

Smushy Circuits

Theory: in order to do work, electricity must flow in a circle back to the battery. Can you test it?

PHYSICAL SCIENCE

You Will Need

- Grown up help - some cooking is required
- A cup of plain flour
- A cup of warm water
- 1/4 cup of salt
- 3 Tbsp Cream of Tartar
- 1 Tbsp oil
- Food colouring - this is optional and added mess!

1/ Mix all the ingredients together in a pot, then heat gently over a low flame until it all clumps together firmly.

2/ Allow to cool. Knead the dough, adding a little extra flour to achieve a good consistency. Careful: too much flour and the dough will crumble, too little and it will stick to your fingers!

3/ Store in an airtight container in a fridge, but for no longer than a week - it will get smelly!

Circuits *must* go in a circle! Make sure there's no short circuits happening anywhere.

Make your own gooey, smushy play dough at www.CreatingScience.Org

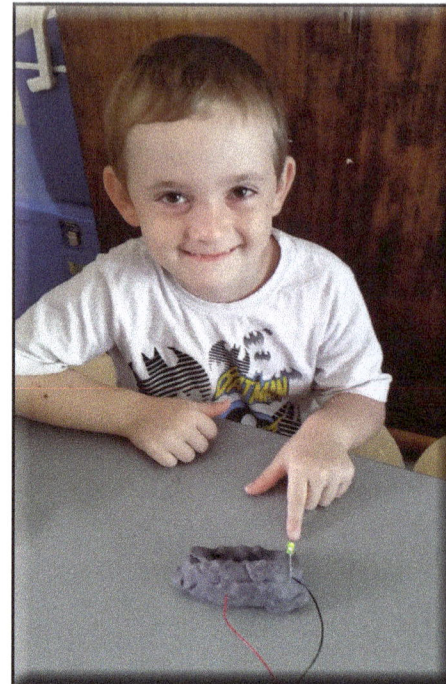

How Does it Work?

Electrical energy flows out of the -ve terminal of the battery, down the black wire, into the short leg of the LED light, through the light to make it glow, down the long leg of the LED, then through the conducting smushy play dough, then finally back down the red wire and into the battery.

So how does the play dough conduct electricity? The salt you put in has an electric charge, and can carry the charge through the dough. Note, however, that it is not really a very good conductor - just a messy one!

Watch out for hot wires! - DO NOT leave your battery attached constantly!

Homopolar Motors

Or 'motors with the electricity going in only one direction' - can you make a motor with magnetism?

Remember: Don't swallow anything in this activity, or anything from this book unless you're told you can!!!

Preparation. You'll need:

Strong magnets, little circle neodymiums are ideal - especially 12mm (circumference) by 5mm (tall).

A short length of conducting wire that ISN'T made out of steel or iron (it will get stuck to your magnet and not move at all). About 10cm will do, make sure the ends are uncovered and clean so the electricity can flow!

A medium sized screw. Nails work; screws are easier to see.

#CreatingScienceHomopolarMotors

Wires carrying electrical currents get hot – sometimes HOT ENOUGH TO BURN FINGERS. Exercise all caution!!!

How to Make a Homopolar Motor:

1. Place the magnet on the flat end of the screw.

2. Place the pointy end of the screw on the battery - it should stay there due to magnetism. If it keeps falling off, get a smaller screw.

3. Hold one end of the wire on the top end of the battery, and gently touch the other end of the wire to the magnet. If electricity is running the magnet will spin!

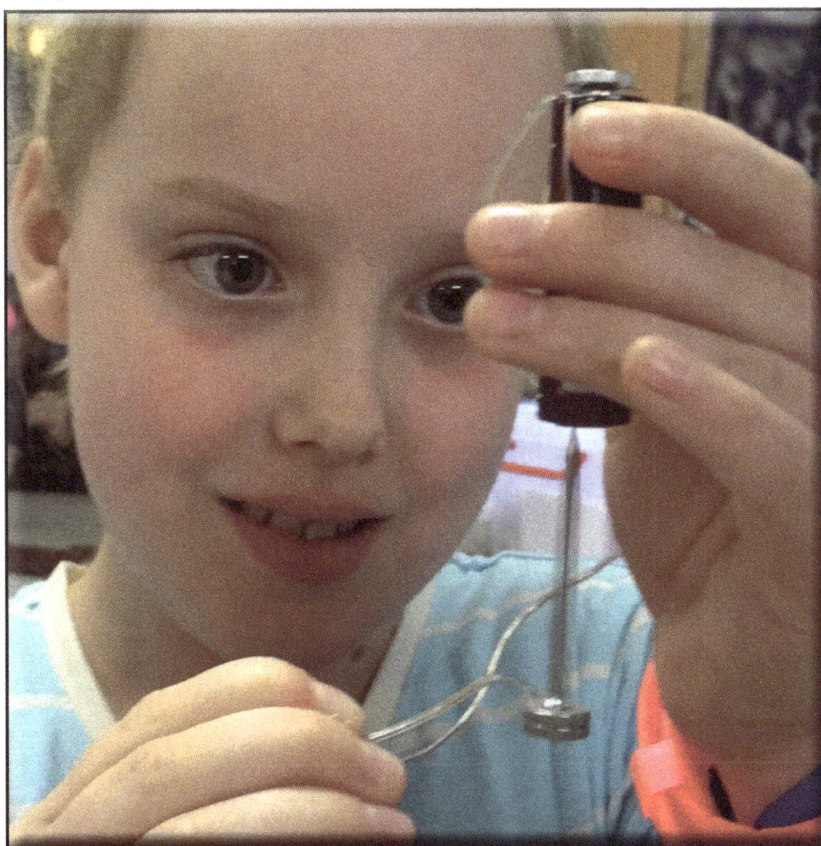

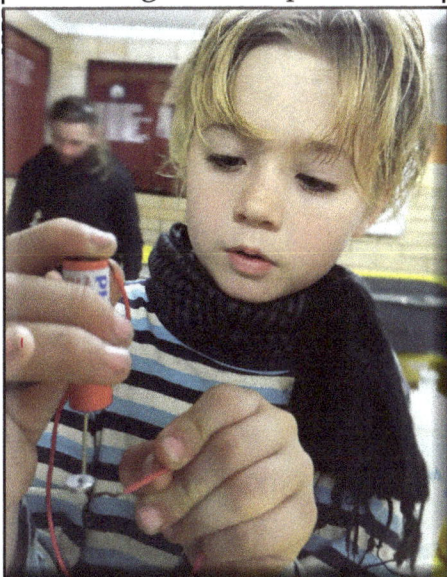

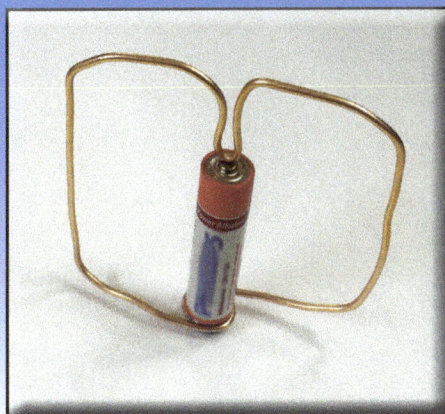

Careful! Wires with electricity can get hot. So why hold on to the wire at all? To make sure it doesn't get so hot it becomes dangerous! Short, ten second bursts are pretty safe.

Look, another successful homopolar motor design!

PHYSICAL SCIENCE

#Creating
Science
ArtBots

Art Bots

Can forces, friction and electricity be used to help us make art? Make a robot that loves to dance and draw!

P H Y S I C A L   S C I E N C E

Warning!

Once again, electricity will make small wires HOT. Be careful!

Materials:

- Small motor with leads
- Three pens with lids on
- Battery pack & batteries
- A strong, light cup with a hole drilled in the top for the battery leads (adult help required!)
- Some decorations, maybe feathers and googly eyes

Feel free to colour and decorate your bot - but remember, the lighter the better!

1. Make sure your offset motor can spin.
(Cat ears not required!!)

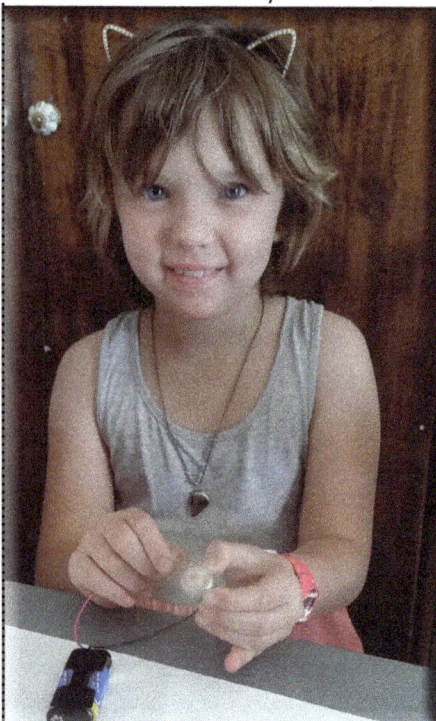

2. Be sure to tape the motor down to the very edge of the cup so that your spinner can spin freely. You might like to use the battery pack as a counterweight.

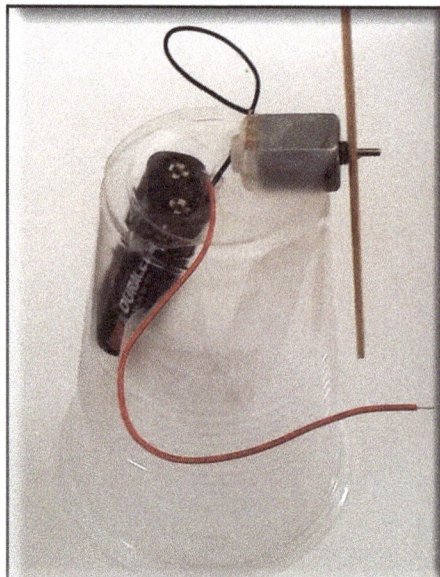

3. Place the three pens (with their lids on for now!) at equal distances around the cup, and tape them down a lot!
You need to make sure your spinner can safely spin without hitting the pens or your fingers! Hook the wires onto the motor and go!

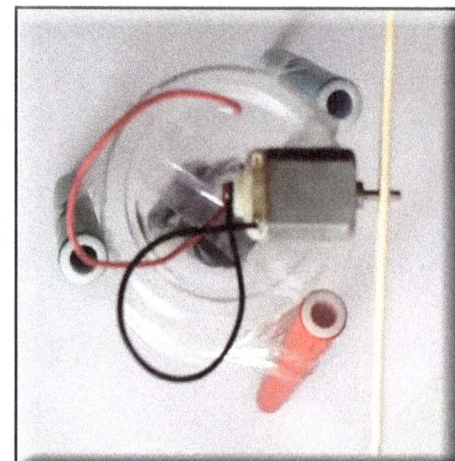

Creating Science!

Earth and Space Science

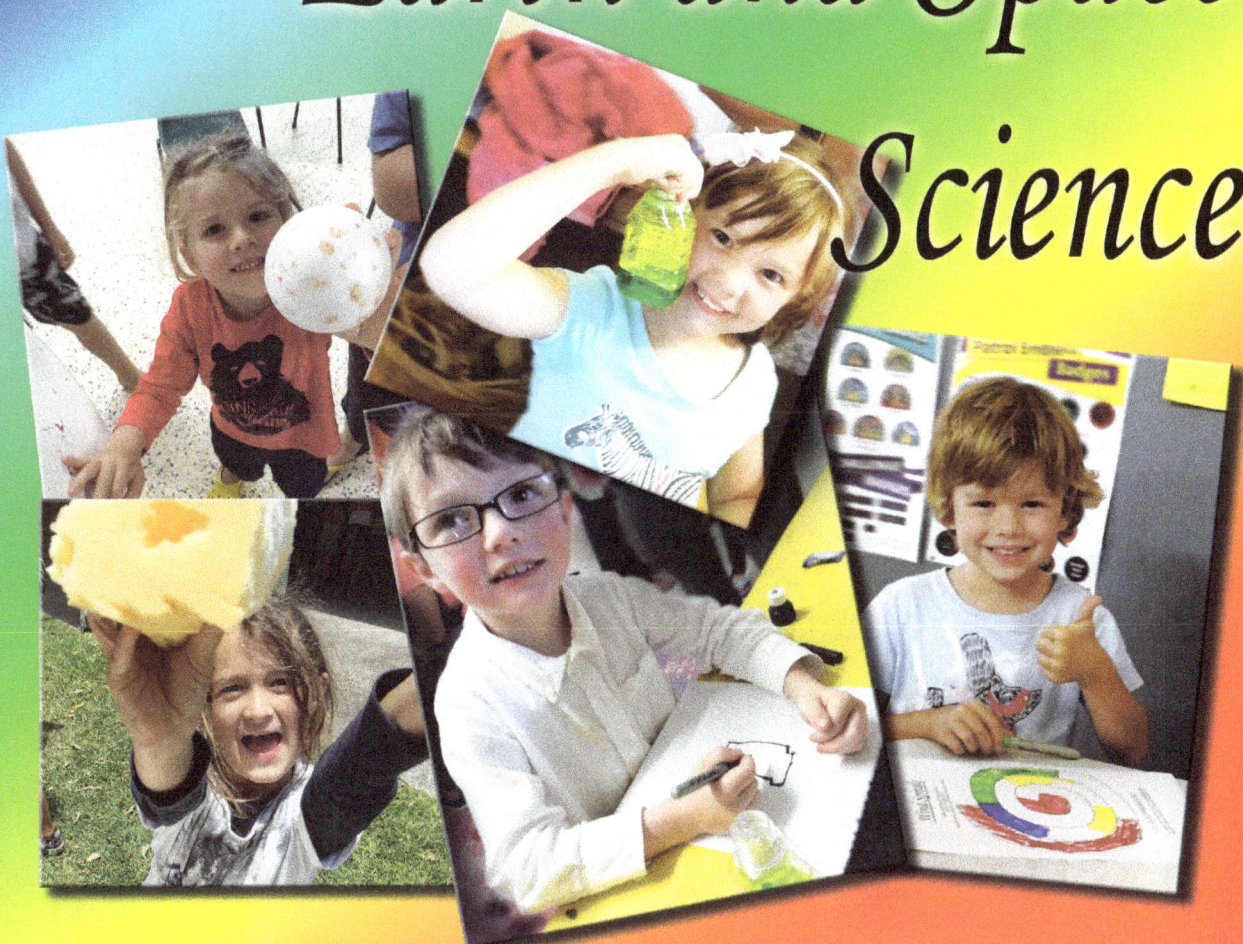

Making Clouds

Can we make a cloud? Indoors!! #CreatingScienceMakingClouds

Let's find out! You need a large, clear plastic bottle - about 2 litres will do - a match, and about ¼ of a cup of water.

EARTH & SPACE SCIENCE

1/ Get your bottle and put in the water. This is the easy part!

2/ Adults: put a lit match in the neck of the bottle & blow it out.

3/ Put the lid on tight, squash the bottle *tight*, and release!

How does it work? We explain this by saying the air has water in it – not very much, just some tiny invisible particles too small to see. When the conditions are just right, some of those particles will gather together to make teeny tiny drops. Suddenly releasing the bottle lowers the pressure, which lowers the temperature, which allows the drops to form. If enough of those teeny tiny drops gather together they'll form a big drop that'll fall right out of the cloud = rain!

Pro Tips: The harder you squeeze, and the faster you let go, the more drops you will form and thus the thicker your cloud will be. Also: the little droplets need somewhere to form on, and that's why we put the smoke from the match in the bottle. Real raindrops often need something like dust to form on as well, and this helps clean up the sky.

Is There Really Water in the Air?

Two ways to demonstrate this: you can fill a cup with icy water and one with room temperature water. After about 10 minutes there should be drops forming on the outside of the icy cup, but not the normal one. The water can't have come through the cup (it didn't on the normal one), so where did the water come from? Perhaps it came from the very air around us? We call the amount of water in the air humidity.

High humidity causes rain, and low humidity contributes to loads of things like cracked lips, more static electricity, and days that seem cooler because sweating is more effective!

Another fun and interesting thing you can do, which ties in well with the page on wind, involves simply placing a clear plastic bag around a leafy branch of a tree. Make sure you tie it well so no air can get in or out. After only one day you can begin to see liquid condensing in the bag and collecting on the bottom. Where did the water come from? Make sure you throw the bag away after a day or the tree will get sick, and don't drink the water, or you may get sick too!

Magma Flows

How do lava and magma differ? Why does magma sometimes flow to the surface of the earth?

Materials:

1)   A clear bottle, preferably taller than wide, about 10cm tall and 3cm wide.

2)   Oil. Vegetable oil is fine.

3)   Drops of food colouring.

4)   Something fizzy. Alka-seltzer is excellent but pricey. Wizz fizz, fruit tingles, vinegar & bicarb also work.

Now for the fun part:

1)   Fill your bottle about 2/3 of the way with oil.

2)   Fill up the rest of the bottle with normal water - which will sink because it is less dense.

3)   Drop in one drop of food colouring.

This is your first observation. What happens to the drop? Why? How long does it take to float down to the bottom and spread out into the water underneath?

Theory: The oil and water are chemically different so that they don't mix. Food colouring is a water based chemical, so it will mix with water and not oil. To mix colouring into oil you need to use oil based colours, or acrylic which can mix with both oil and water.

4)   Now drop in your fizzy thing.

Next observation: What happens when you drop in the fizzy thing, and why?

Theory: The ingredients in the fizzy thing react with the water (and not the oil) to produce carbon dioxide gas. This gas has a strange yet entertaining effect on the coloured water. Usually water is more dense than the oil, so it stays at the bottom. But when the coloured water collects a few bubbles its total density is now less than the oil. So it floats to the top.

Third observation: Once the coloured drops get to the top, what happens?

Theory: The coloured water rises to the top due to the little bubbles stuck to it. The bubbles then pop once they reach the air. Thus, the water drops no longer have the air bubbles to help keep their density lower than oil, and they sink back down.

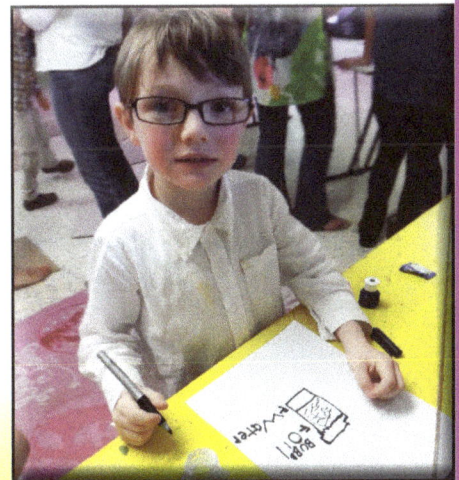

How is This Like Lava?

Modelling: The magma bottle acts like real magma under the surface of the earth. It too has lots of gas trapped inside it. When magma is forced up to the surface of the earth it, too, can explode sending literally tonnes of extra gas outwards into the air, including carbon dioxide, sulphates, and water. The magma has now chemically changed, and we call it lava. The difference between our model and the earth's crust is that the lava can take millions of years to cycle back down. Lava tends to stay on the surface of the earth, forming new land or giant volcanoes. It won't recycle back down to form new land until millions of years have passed, or perhaps forever…

EARTH & SPACE SCIENCE

Wind Cannon

Air is around us all the time, so what makes it move?

EARTH & SPACE SCIENCE

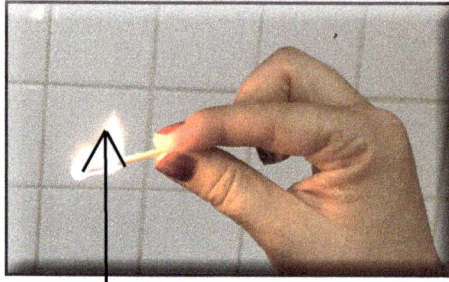

Hot air rises

All air, everywhere, is pushing in all directions all the time. The further down you go, the more the air will push and squeeze together. When hot air pushes around, it finds the weakest point directly above it, so up it goes. When cold air pushes around, it finds the easiest way is to burrow down through the surrounding air, because it is much more dense. In both cases the air nearby will push sideways to get into the gap left by the other air - creating wind!

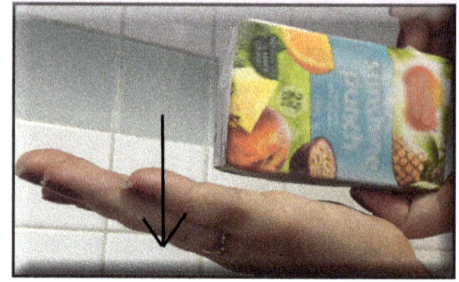

Cold air falls

Make Your Own Slightly Dangerous Air Pressure Wind Cannon

You will need two balloons, scissors, sticky tape, and a cardboard roll - toilet rolls work fine! Now, with grownup help, carefully cut the tip off a balloon. Then cut the nozzle off that balloon and one other balloon.

Stretch the cut ends of the balloons over a small cardboard roll. Tape them thoroughly in place.

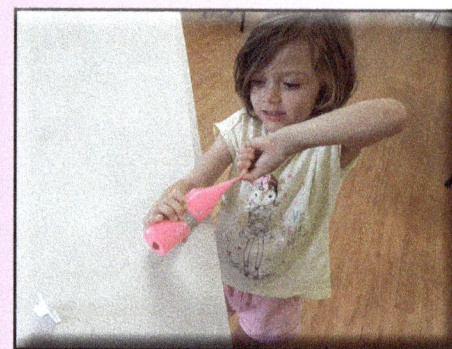

Carefully pull the balloon without the hole backwards, and let it go suddenly to increase the pressure in the tube so much that it shoots a puff of wind out of the end with a hole. Can you use the wind to knock over paper?

Air is either warmed up, or cooled down, by lots of different things - a hot pavement, a cold sea, a tall mountain, or a room full of excited kids!

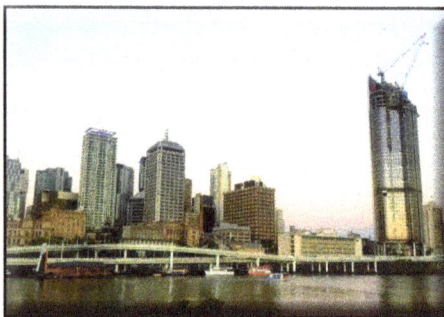

A warm city with lots of concrete will heat up the nearby air, making it rise.

Water cools down *and* heats up slower than the land, causing the sea breeze.

Air is also cooled as it rises up, so high mountains can make air very cold.

#Creating Science Vortexes

Vortexes

Tornadoes, the weather, the sun ... vortex motion is everywhere! Can you make your own vortex?

Build your own vortex bottle with our special connector and two soft drink bottles. Holding the top and middle, turn the bottles water side up and give it a spin. Check it out!

If you spin it hard enough, and wait a moment, you should see the water swirl around in a cyclone shape. We call this shape a 'Vortex'. Why? The water is flowing down and the air is pushing up, and in order to do this as easily as possible, they will co-operate. This is what creates the vortex.

Not technical enough? Try: Both water and air have 'pressure', which means they are both pushing outwards in all directions. (So the air in the bubbles is pushing the water away, otherwise, you could not see the bubble at all! And air has a LOT of push.) The water pressure is also pushing in on the bubbles, but as the water spins it loses pressure in the middle, so the bubbles all join together and push their way through!

Vortexes happen all the time in nature. For instance, when you run your hand through air or water, tiny little vortexes form off the leading edges.

Note: Every vortex has a low pressure area in the middle, and the fluid is spinning around that point. At the Earth's poles there is a year-round vortex of low pressure cold air. German scientist and philosopher Hermann von Helmholtz was probably the first to investigate the properties of vortex motion in the mid 1800s, though even Leonardo DaVinci in the 1500s studied fluid motion and speculated on vortexes.

Remember, air never sucks, it only ever pushes. So you don't get sucked into tornadoes or whirlpools, but pushed from behind by the materials that are rushing in!!

The Coriolis Effect

Some people think that since the weather turns clockwise in the southern hemisphere, and anticlockwise in the northern, that water going down the sink and toilet must as well. But research shows you need to be *big* for the turning of the earth to affect which way you spin. As big, for example, as the *weather*. Sinks and toilets are just too small!!

A water vortex dimple on the water, with a vortex shadow underneath where the curved water acts like a lens to light.

EARTH & SPACE SCIENCE

Volcanoes

So important, so destructive! Can you make your own improved model volcano?

E A R T H & S P A C E S C I E N C E

1. Grab an old foam mattress. Rip or cut it into a rough volcano shape.

2. Make sure you dig a caldera in the top. Keep the sides sloped like a volcano.

3. You can even paint it all over to help make it look more realistic!

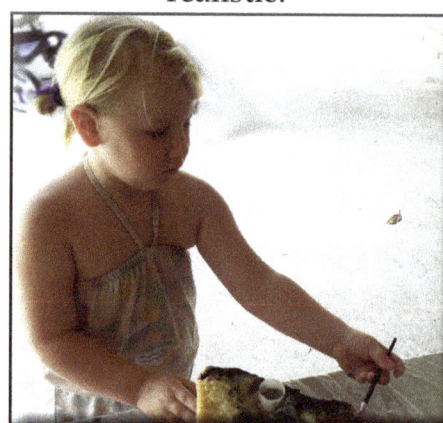

Improved Volcano Lava

We all know about using bicarb and vinegar to make pretend lava, but di you know you can:

1. Add food dye to make your lava look extra red.

2. Add dishwashing detergent for long lasting lava and a convincing pyroclastic flow.

3. (Explosion - grownup and goggles required!) Put your bicarb in a tissue, inside a tiny sandwich ziploc bag that has a little bit of vinegar, and BOOM!

STAY AWAY: How Volcanoes Can Harm You:

The lava can incinerate you, it is 4000°C.

The pyroclastic flow: Super-heated particles such as sand flowing down like an avalanche can burn forests in seconds.

A single large volcano can change the climate worldwide (and may have helped kill the dinosaurs).

Volcanoes produce acid rain.

The gasses can suffocate you, including carbon dioxide.

Airplane crashes - Ash thrown into the air by eruptions can present a hazard to aircraft.

You can get hit by flying rocks.

#CreatingScienceVolcanoes for more!

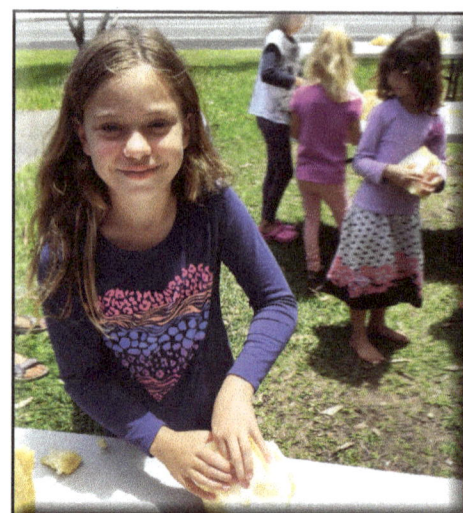

Growing Crystals

Warning: This activity uses boiling water – adult help and care required!

Igneous rocks are "born from fire" - or in our case, hot water.

You will need:

- A metal salt, such as copper sulphate (although table salt will also do fine!)
- Boiling hot water.
- String, ~30cm, and a stick to tie it on to.
- Somewhere to keep the growing rocks, such as a small glass or plastic bottle.
- A crystal seed, see below.

1. Pour the boiling water into your container. Tie the string to your stick so that it almost touches the bottom.

2. Add your salt and stir – keep adding till no more salt will dissolve! (It's called 'super saturating'.)

3. Place your string hanging freely in the bottle and leave it, usually for several weeks!

Why it works: As the water evaporates, the left over salt forms a solid – usually a cube of positive and negative particles packed next to each other like bricks in a wall. If this reaction happens slowly, the particle 'bricks' (called *ions* because they have electrical changes) have time to pack neatly, forming large, impressive crystals of different shapes for different chemicals!

If the reaction happens quickly (which can be anything from minutes to centuries, depending on the materials being used) the crystals form erratically. This results in countless small crystals growing in erratic directions, often looking like a lumpy mass that can still be hard as concrete. Actually, that's a good example of a fast setting material with irregular crystals – concrete!

Advice:

1. It helps, it really does, to have a 'crystal seed' attached to the string before you begin. You can do the activity on a flat plate first, tying the largest, neatest crystal to the string as a 'seed' to grow more impressive crystals.

2. Keep it clean - even colouring can act as pollution, resulting in smaller crystals.

3. Make even larger crystals by renewing the solution every few days with more super-saturated solution.

4. The SLOWER the reaction occurs, the larger and less messy the crystals tend to be. Can you see the top of the string form irregular crystals where the water evaporates quite quickly?

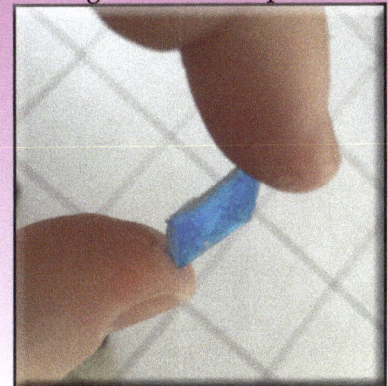

A crystal seed grown in an open cup

EARTH & SPACE SCIENCE

Sedimentary Sculptures

Sedimentary rocks grow over time as sand and other sediments gather. Can you make a dino graveyard?

(Watch out for scratchy sand!) #CreatingScienceSedimentarySculptures

EARTH & SPACE SCIENCE

Top layer: plants, soil etc

Bedrocks

Bore water

Oil and coal

Complex life

(A fossil!)

Early life

Pre-life layer

Magma layer

Did You Know the Earth Grows in Layers?

The ground you walk on is, usually very slowly, changing. Volcanoes and floods can leave new layers of earth. Plants turn into soil, and the soil can eventually turn into stone. Dust even falls in from outer space! It may take millions of years for these layers, called sediments, to form. Digging down is like digging into the past to see what was around back then!

Try This Simple and Fun Activity to Build Your Own 'Sedimentary Layer Bottle'

- A small, clear bottle. Glass or plastic will do, but it's helpful if it has a screw top.
- Clean sand, the whiter the better.
- Acrylic paints (food dye or printer ink also work) to colour the sand.
- Zip lock bags to mix the sand. A spoon and funnel to add sand to the bottle. You may need tissue paper and sticky tape to help press the sand down tightly into the bottle and keep it there! With lots of sand to work with it's a good idea to work outside.

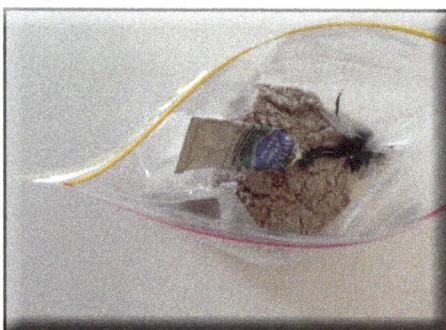

1. Put about half a cup of sand in a zip lock bag, along with a small amount of paint or dye. Close the bag tight!

2. Mix thoroughly. Open it up and allow to dry - spread out on baking paper if needed. Repeat for different colours.

3. Using a funnel and a spoon, place one colour at a time to create 'sedimentary layers'.

#Creating
Science
TheMoon

The Moon

So much to learn about Earth's nearest neighbour! What shape is the moon? What is it made from?

A near side and...

A far side!

The Moon Has Two Sides!

One side always faces the Earth. Long ago, when the moon was forming, lava would sometimes flow to its surface. Just like the tides on earth, the tides on the moon made the near side of the moon have large 'seas' of lava, which have turned into stone now.

The far side never faces earth, and is covered in craters. Sometimes the far side is called the 'dark side of the moon' - which is wrong, it isn't any darker than the near side, it just never faces toward planet Earth.

How Old is The Moon?

Scientists currently think that around four and a half billion years ago, when the Earth was still forming and would have looked like a ball of molten lava, another young planet smashed into it. Much of the 'splash' from this collision gathered together to form the moon, and the rest stuck together to become the Earth we know and love. Evidence includes the fact that the rocks on both places are around about the same age, and chemically more or less the same - what do you think?

Craters!

Make your own craters by dropping (not throwing) stones into sand that is covered with flour. Note the *ejector trails* where large amounts of material come flying out of the ground to make the *crater*.

The faster and heavier a meteorite is, the bigger the crater will be.

EARTH & SPACE SCIENCE

The Planets

How big, and how far, would the planets be if YOU were the sun?

EARTH & SPACE SCIENCE

The size of things in space is really amazing!

Imagine the sun was a ball about as big as you are (about 1.4 meters tall). You can even draw a big circle out on the ground in chalk to get a sense of how big that might be.

So tell me, how big would the Earth be, if the sun was a big ball about as big as you?

Can you guess?

Look on the table in the appendix to find the relative sizes and distances of the planets, as well as something special about each one.

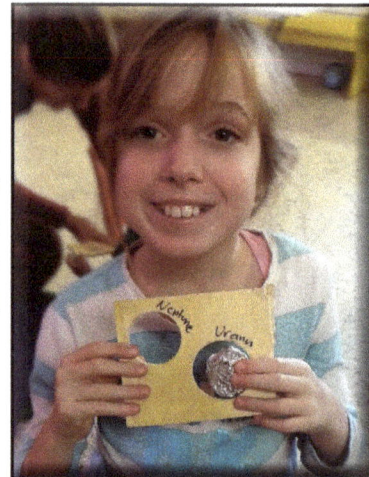

Full size images @ #CreatingScienceThePlanets

How would you arrange the planets? In groups according to size? Colour? Interesting features? For a long time, we said there were 9 planets: 4 rocky inner planets, 4 gas planets, and 1 dirty snowball called Pluto. But scientists realised that, if the maths was correct, there could be literally *hundreds* of dirty snowballs out beyond Neptune. Then… they found one…

It was called UB313 2003 at the time, as it waited for an official name. In the meantime the debate started: do we have 10 planets in the system? What if there were more dirty snowballs? Did we really want 50+ planets in the solar system,

which is what would happen if we kept finding planets. Sure enough, two weeks later, two more dirty snowballs were found. It was happening… 12 planets, or even more?

Scientists decided it was time to sit down and decide what the name 'planet' officially meant. In the end, they decided:

1. A planet has to go around a sun, not another planet.

2. A planet must be big enough to be round (so comets are not planets).

3. A planet has to dominate its orbit around the sun, so it's not sharing it (like Ceres in the asteroid belt is).

And this meant that all the dirty snowballs like Pluto, many of which aren't round and none of which dominate their own orbit, were no longer classified as planets. Now, they are classified as **dwarf planets**. And the dwarf planet UB313 2003 was called Eris, after the Greek goddess of *arguments*.

Do you think Pluto and Eris should be called planets?

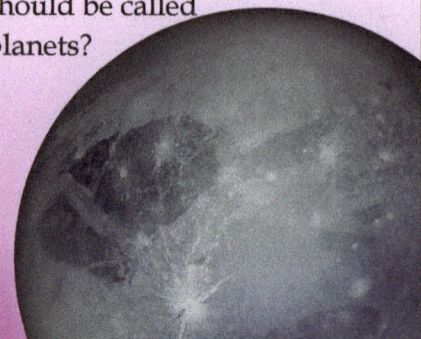

This page may be copied for incidental educative purposes only   (c) *Dr Joseph Ireland 2018*

How BIG are the planets

If YOU were the sun!

Mercury

Venus

Earth

Mars

Pluto

Eris

Jupiter

Uranus

Saturn

Neptune

EARTH & SPACE SCIENCE

Solar System Facts

EARTH & SPACE SCIENCE

Name	Size	Distance (m)	Notes
The sun	You (1.4 meters around)	Right here	Hot, really, really hot! At least 4000 degrees at its surface, and over a million in its enormous atmosphere. Is also the solar system's most powerful magnet.
Mercury	0.48 cm - a small pebble	58	Mercury is the smallest planet, but still not too small: it's as big as Australia! It's also the fastest planet, only 88 days in a year. With almost no air it is boiling hot in the sun, and freezing at night!
Venus	1.21cm - a fat pea	108	Venus is the hottest planet, 500ºC all day and all night. It is about as big as Earth. It rotates so slowly its day is longer than its 288 earth day year!
Earth	1.28cm - a fatter pea	150	Earth is the only planet with life on it, so far as we know... Our moon is bigger than Pluto, but smaller than Mercury. Earth has the third largest magnetic field in the solar system.
Mars	0.68cm - a beanbag bean	228	Mars is red due to iron in the dust, like outback Australia, but it's not very warm: 0 degrees Centigrade is an average day! It's also not very big, so gravity is low. It's only as wide as the Andes mountain range (South America), or a big ball from Brisbane to Hong Kong.
Ceres and the asteroid belt	.07cm - a mote of dust	420	Ceres, the largest asteroid of the asteroid belt (about half way between Jupiter and Mars) is about the size of Victoria, but it isn't counted as a planet... do you think it should be? Also, most movies make out asteroid belts to be a crowded, deadly place. Actually most are just grains of dust, and they're so far apart from each other that it's almost impossible to see one from the next.
Jupiter	14.30cm - a small ball	778	The first of the four gas planets, Jupiter is the largest planet: bigger than the rest put together! The great red spot, larger than 3 Earths side by side, is a violent storm that's being going for at least 300 years. Jupiter has the second strongest magnetic field in the solar system, after the sun.
Saturn	12.05cm - a vinyl record with a ball in the middle	1,427	Saturn's spectacular rings are as wide as five Earths, and yet so thin you could walk from the top to bottom, if you could walk in space (they are only a kilometer thick sometimes). All the Gas planets have rings. The largest moon in the solar system is Titan of Saturn - it is bigger than Mercury!
Uranus	5.11cm - a handball ball	2,871	The first of the Ice Giants, Uranus spins almost on its side... so one side gets all the sun in 'summer', day and night! The astronomer who found it wanted to call it George, after the king. But the International Astronomical Society voted no.
Neptune	4.95cm - an undersized handball	4,498	Neptune is the last planet in our solar system - so far as we know for now! It also has the fastest winds of any planet, at times more than 2000 km per hour!
Pluto	0.23cm a really small grain of sand	5,900	No longer considered a planet, Pluto is now called a dwarf planet. Even so, it's not that small, it's about the size of Queensland. Pluto is considered the first of the TNO – Trans Neptunian Objects. It is cratered, like our moon, but MUCH colder!
Eris	0.23cm a really small but heavier than Pluto grain of sand	14,600 meters or 14.6 kilometres	Smaller but heavier than Pluto, finding Eris sparked the debate that demoted Pluto to dwarf planet status. Eris is named after the goddess of arguments. Eris is the largest TNO, which are all made of ice and dust, not gas or rock. There are loads of other TNO's, maybe more than we can ever count! But some big ones include Haumea, Make Make, and Sedna, but scientists are still finding more!
Heliopause	(A big ball 24 kilometres across)	Around 24 kilometers away	Where the solar wind is balanced out by the local galactic climate. Is this the true end of the Solar System? Some would disagree!

Sundial

55 50 45 40 35 30 25

2. Fold up here along the line that best describes your current latitude.

1. Cut here & 3. Tape down

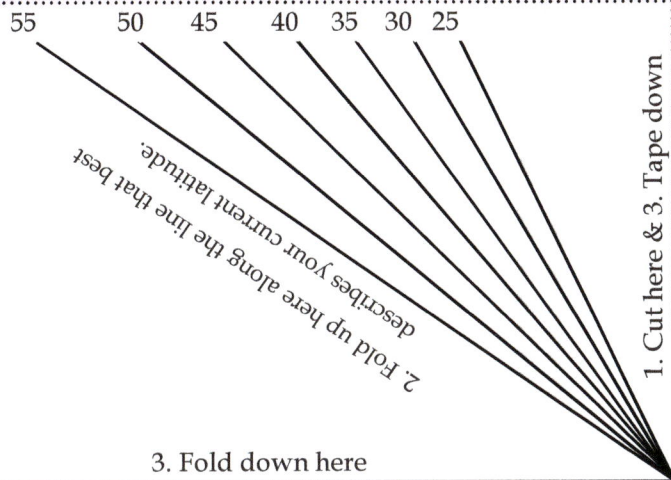

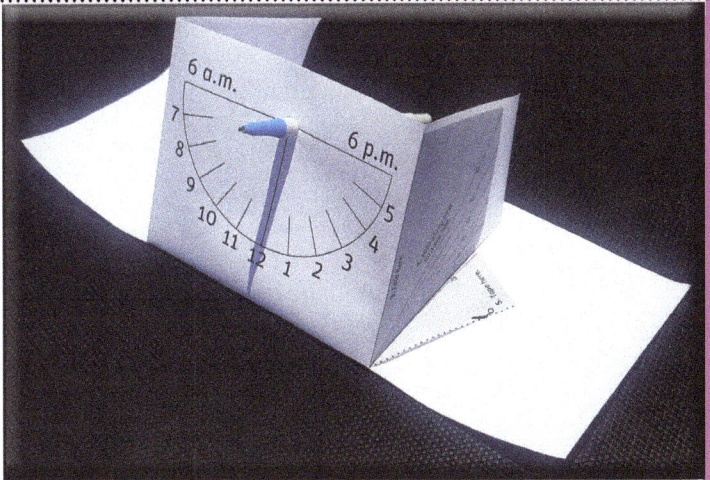

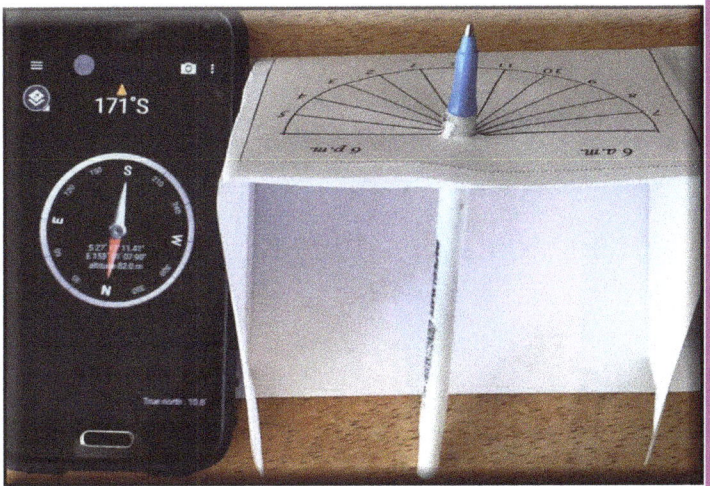

3. Fold down here

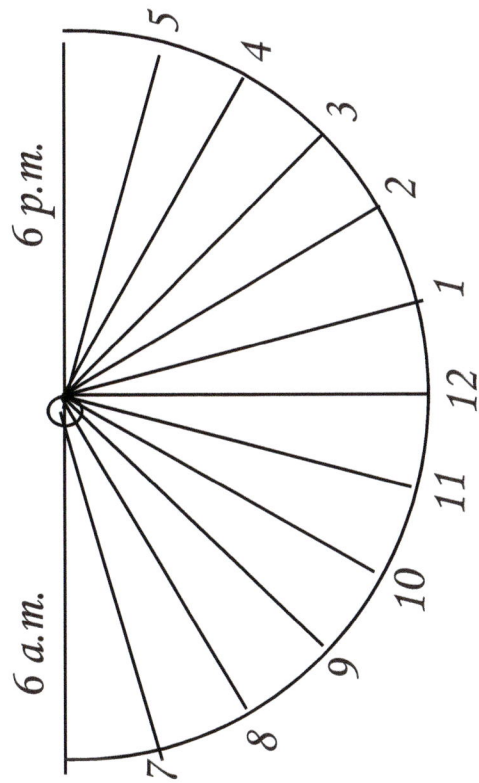

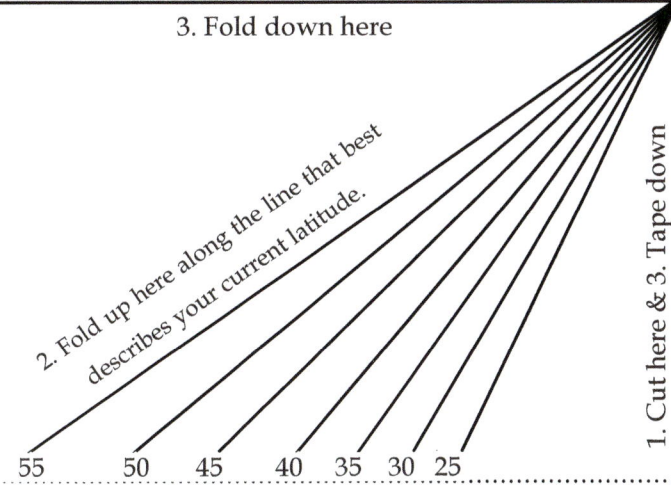

5
4
3
2
1
12
11
10
9
8
7

6 p.m.

6 a.m.

3. Fold down here

2. Fold up here along the line that best describes your current latitude.

55 50 45 40 35 30 25

1. Cut here & 3. Tape down

A Southern Hemisphere Sundial.

1. Copy this page. Cut along the dotted lines.

2. Research your latitude on the Earth, and fold the two sides depending on it as shown.

3. Then fold, arrange and tape together the sundial as shown in the diagrams.

4. Push a straight pen or pencil through the central circle as shown.

5. Align your pencil pointing directly towards the nearest pole (south in the southern hemisphere, for example). Set up in the sun light.

Tape it down if you need some stability. Remember, if your local community has daylight savings you'll need to take an hour away during summer. Your finished sundial might be a little rough, but it's still amazing: The dial itself is actually parallel with the equator of the Earth!

Download online at #CreatingScienceSundial

EARTH & SPACE SCIENCE

An extra big thanks to all who made it this far.

If you'd like to try more activities and experiments, don't forget to check out our webpage:

www.CreatingScience.Org

Find us anywhere online with the hashtag

#CreatingScience

Followed by the chapter title.

For example:

#CreatingScienceBobTheBlob or
#CreatingScienceRocketBalloons

Share experiences

Give feedback

Find and offer helpful suggestions

Tell us what works, and what doesn't

Share the fun!

Dr. Joe
the Travelling Scientist

Proudly presented by Creating Science

www.ingramcontent.com/pod-product-compliance
Lightning Source LLC
Chambersburg PA
CBHW040301100426
42811CB00011B/1332